T013447

Lecture Notes in Electrical Engineering

Volume 470

** Indexing: The books of this series are submitted to ISI Proceedings, EI-Compendex, SCOPUS, MetaPress, Springerlink **

Lecture Notes in Electrical Engineering (LNEE) is a book series which reports the latest research and developments in Electrical Engineering, namely:

- Communication, Networks, and Information Theory
- Computer Engineering
- Signal, Image, Speech and Information Processing
- Circuits and Systems
- Bioengineering
- Engineering

The audience for the books in LNEE consists of advanced level students, researchers, and industry professionals working at the forefront of their fields. Much like Springer's other Lecture Notes series, LNEE will be distributed through Springer's print and electronic publishing channels.

For general information about this series, comments or suggestions, please use the contact address under "service for this series".

To submit a proposal or request further information, please contact the appropriate Springer Publishing Editors:

Asia:

China, *Jessie Guo, Assistant Editor* (jessie.guo@springer.com) (Engineering)

India, *Swati Meherishi, Senior Editor* (swati.meherishi@springer.com) (Engineering)

Japan, *Takeyuki Yonezawa, Editorial Director* (takeyuki.yonezawa@springer.com) (Physical Sciences & Engineering)

South Korea, *Smith (Ahram) Chae, Associate Editor* (smith.chae@springer.com) (Physical Sciences & Engineering)

Southeast Asia, *Ramesh Premnath, Editor* (ramesh.premnath@springer.com) (Electrical Engineering)

South Asia, *Aninda Bose, Editor* (aninda.bose@springer.com) (Electrical Engineering)

Europe:

Leontina Di Cecco, Editor (Leontina.dicecco@springer.com)
(Applied Sciences and Engineering; Bio-Inspired Robotics, Medical Robotics, Bioengineering; Computational Methods & Models in Science, Medicine and Technology; Soft Computing; Philosophy of Modern Science and Technologies; Mechanical Engineering; Ocean and Naval Engineering; Water Management & Technology)

(christoph.baumann@springer.com)
(Heat and Mass Transfer, Signal Processing and Telecommunications, and Solid and Fluid Mechanics, and Engineering Materials)

North America:

Michael Luby, Editor (michael.luby@springer.com) (Mechanics; Materials)

More information about this series at http://www.springer.com/series/7818

Jaydeb Bhaumik · Indrajit Chakrabarti
Bishnu Prasad De · Banibrata Bag
Surajit Mukherjee
Editors

Communication, Devices, and Computing

Proceedings of ICCDC 2017

 Springer

Editors

Jaydeb Bhaumik
Department of Electronics and
    Communication Engineering
Haldia Institute of Technology
Haldia, West Bengal
India

Banibrata Bag
Department of Electronics and
    Communication Engineering
Haldia Institute of Technology
Haldia, West Bengal
India

Indrajit Chakrabarti
Department of Electronics and Electrical
    Communication Engineering
Indian Institute of Technology Kharagpur
Kharagpur
India

Surajit Mukherjee
Department of Electronics and
    Communication Engineering
Haldia Institute of Technology
Haldia, West Bengal
India

Bishnu Prasad De
Department of Electronics and
    Communication Engineering
Haldia Institute of Technology
Haldia, West Bengal
India

ISSN 1876-1100   ISSN 1876-1119   (electronic)
Lecture Notes in Electrical Engineering
ISBN 978-981-13-4191-5   ISBN 978-981-10-8585-7   (eBook)
https://doi.org/10.1007/978-981-10-8585-7

Printed on acid-free paper

This Springer imprint is published by the registered company Springer Nature Singapore Pte Ltd.
part of Springer Nature
The registered company address is: 152 Beach Road, #21-01/04 Gateway East, Singapore 189721, Singapore

Preface

The first International Conference on *Communication, Devices, and Computing* (ICCDC 2017) was held at Haldia Institute of Technology, Haldia, during November 2–3, 2017. Haldia is a city in Purba Medinipur district of West Bengal, India. Haldia Institute of Technology is dedicated to the objectives of creating highly trained professional manpower in various disciplines of engineering. It has gained a reputation through institutional dedication to teaching and research.

In response to call for papers of ICCDC 2017, a total of 62 papers were submitted for presentation and inclusion in the proceedings of the conference. These papers were evaluated and ranked on the basis of their novelty, significance, and technical quality by at least two reviewers per paper. After a careful and blind refereeing process, 29 papers were selected for inclusion in the proceedings. These papers cover current research in communication, signal processing, image processing, wireless network, semiconductor devices, VLSI design, antenna design, and fuzzy modeling. The conference hosted seven invited talks by Prof. Animesh Maitra (Calcutta University), Prof. Amlan Chakrabarti (Calcutta University), Prof. Sukumar Nandi (IIT Guwahati), Prof. Indrajit Chakrabarti (IIT Kharagpur), Prof. Susanta Kr. Parui (IIEST, Shibpur), Prof. Manoranjan Maiti (Vidyasagar University), and Mr. Aninda Bose (Senior Editor, Springer).

A conference of this kind would not be possible without the full support from different committee members. The organizational aspects were looked after by the organizing committee members who spent their time and energy in making the conference a reality. We also thank all the technical program committee members and additional reviewers for thoroughly reviewing the papers submitted to the conference and sending their constructive suggestions to improve the quality of papers. Our hearty thanks to Springer for agreeing to publish the conference proceedings.

We are indebted to DRDO, ISRO, IGI Global, IETE, and Haldia Institute of Technology for sponsoring and supporting the event. Last but not least, our sincere thanks go to all speakers, participants, and all authors who have submitted papers to ICCDC 2017. We sincerely hope that the readers will find the proceedings stimulating and inspiring.

Haldia, India   Jaydeb Bhaumik
Kharagpur, India   Indrajit Chakrabarti
Haldia, India   Bishnu Prasad De
Haldia, India   Banibrata Bag
Haldia, India   Surajit Mukherjee

Committee

Patron

Lakshman Seth, Chairman, Haldia Institute of Technology, India

Joint Organizing Secretary

Jaydeb Bhaumik, Head, Department of ECE, Haldia Institute of Technology, India
Bishnu Prasad De, Haldia Institute of Technology, India

Joint Convener

Surajit Mukherjee, Haldia Institute of Technology, India
Banibrata Bag, Haldia Institute of Technology, India

Program Committee Members

PC Members, Organization

Bidyut B. Chaudhuri, Indian Statistical Institute, Kolkata
Santanu Chaudhury, Director, CEERI, Pilani, and Professor, IIT Delhi
Sankar K. Pal, ISI, Kolkata
S. S. Pathak, IIT Kharagpur
Sukumar Nandi, IIT Guwahati
Animesh Maitra, University of Calcutta
Indrajit Chakrabarti, IIT Kharagpur

Ranjan Ganguli, Indian Institute of Science, Bangalore
Amlan J. Pal, Indian Association for the Cultivation of Science (IACS), Kolkata
Chandan Kr. Sarkar, Jadavpur University, Kolkata
Anindya Bose, Burdwan University, Burdwan
Shaurya Agarwal, California State University, Los Angeles
Amlan Chatterjee, California State University, Dominguez Hills
Asit Baran Mandal, CSIR-CGCRI, Kolkata
Md. Mostafizur Rahman, Khulna University of Engineering and Technology (KUET)
Kalyanmoy Deb, Michigan State University, USA
Hiroyuki Tsuda, Keio University, Japan
Jari Kaivo-oja, Research Director, Finland Futures Research Centre, TSE, University of Turku, Finland
Santi P. Maity, IIEST, Shibpur
Sajal Biring, Ming Chi University of Technology, Taiwan
AKS Chandele, President, IETE, PVSM, AVSM (Retd)
KTV Reddy, Chairman, TPPC, IETE
Pabitra K. Chakrabarti, University of Burdwan, Burdwan
Mohiuddin Ahmad, KUET, Khulna, Bangladesh
Manik Lal Das, Professor, DA-IICT, Gujarat, India

Technical Program Committee Members

TCP Members, Organization

Sumit Kundu, NIT Durgapur
Amlan Chakrabarti, University of Calcutta, Kolkata
Debaprasad Das, Assam University, Silchar
Nisha Gupta, Birla Institute of Technology, Mesra, Ranchi
Dulal Acharjee, Applied Computer Technology, Kolkata
Malay Kumar Pandit, HIT, Haldia
Susanta Kumar Parui, IIEST, Shibpur
Ananda Shankar Chowdhury, JU, Kolkata
Anjan Chakravorty, Indian Institute of Technology, Madras
Angsuman Sarkar, Kalyani Government Engineering College, Kalyani
Bubu Bhuyan, North-Eastern Hill University, Shillong
Rajarshi Mahapatra, IIIT Naya Raipur, Chhattisgarh
Sunandan Bhunia, CIT, Kokrajhar, Assam
Manodipan Sahoo, HIT, Haldia
T. Satyanarayana, LBRCE, Andhra Pradesh
Santosh Biswas, IIT Guwahati
Chinmoy Saha, IISST, Thiruvananthapuram
Sudipta Chattopadhyay, Mizoram University, Mizoram

Additional Reviewers

Reviewer, Organization
Asit Baran Maity, HIT, Haldia
Uday Maji, HIT, Haldia
Bidyut Das, HIT, Haldia
Dilip Kr. De, HIT, Haldia
Pabitra Pal, Vidyasagar University
Supriyo De, Saroj Mohan Institute of Technology
Soumen Paul, HIT, Haldia
Chanchal Kr. De, HIT, Haldia

Advisory Committee

Sayantan Seth, Vice Chairman, HIT Haldia
Asish Lahiri, Secretary, HIT Haldia
M. N. Bandyopadhyay, Director, HIT, Haldia
A. K. Saha, Principal, HIT, Haldia
Anjan Mishra, Registrar, HIT, Haldia
Asit Baran Maity, Dean, SASH, HIT, Haldia
Debasis Giri, Dean, School of ECI, HIT, Haldia
S. K. Basu, Finance Manager, HIT, Haldia
Debasis Das, Sr. Administrative Officer, HIT, Haldia

Organizing Committee

Tilak Mukherjee
Jagannath Samanta
Santanu Maity
Dipak Samanta
Kushal Roy
Avisankar Roy
Pinaki Satpathy
Jayanta Kr. Bag
Dipak Samanta
Razia Sultana
Sayani Ghosh
Pulak Maity
Subhendu Barman
Atanu Pradhan

Ira Samanta
Tanushree Bera
Asim Kumar Jana
Wriddhi Bhowmik
Amit Bhattacharyya
Suman Paul
Raj Kumar Maity
Tirthadip Sinha
Akinchan Das
Dibyendu Chowdhury
Avishek Das
Moumita Jana
Sourav Kr. Das
Sachindeb Jana
Tapan Maity
Asim Kuilya
Shampa Biswas Samanta

Contents

Binary Error Correcting Code for DNA Databank 1
Jagannath Samanta, Jaydeb Bhaumik, Soma Barman and Raj Kumar Maity

**Proactive and Reactive DF Relaying for Energy Harvesting Underlay
CR Network** 13
Mousam Chatterjee, Subhra Shankha Bhattacherjee
and Chanchal Kumar De

**Butler Matrix Fed Exponentially Tapered H-Plane Horn Antenna
Array System Using Substrate Integrated Folded Waveguide
Technology** 25
Wriddhi Bhowmik, Vibha Rani Gupta, Shweta Srivastava
and Laxman Prasad

**Computing Characteristic Impedance of MIM Nano Surface Plasmon
Structure from Propagation Vector Characteristics for Skin Depth
Measurement** 39
Pratibha Verma and Arpan Deyasi

Extended Directional IPVO for Reversible Data Hiding Scheme 47
Sudipta Meikap and Biswapati Jana

**Hamming Code-Based Watermarking Scheme for Image
Authentication and Tampered Detection** 59
Pabitra Pal, Partha Chowdhuri, Biswapati Jana and Jaydeb Bhaumik

**RS (255, 249) Codec Based on All Primitive Polynomials
Over GF(2^8)** 69
Jagannath Samanta, Jaydeb Bhaumik, Soma Barman, Sk. G. S. Hossain,
Mandira Sahu and Subrata Dutta

**Secure User Authentication System Using Image-Based OTP and
Randomize Numeric OTP Based on User Unique Biometric Image and
Digit Repositioning Scheme**   83
Ramkrishna Das, Sarbajit Manna and Saurabh Dutta

**Application of RCGA in Optimization of Return Loss of a Monopole
Antenna with Sierpinski Fractal Geometry**   95
Ankan Bhattacharya, Bappadittya Roy, Shashibhushan Vinit
and Anup K. Bhattacharjee

**Improvement of Radiation Performances of Butler Matrix-Fed
Antenna Array System Using 4 × 1 Planar Circular EBG Units** 103
Wriddhi Bhowmik, Surajit Mukherjee, Vibha Rani Gupta,
Shweta Srivastava and Laxman Prasad

**Improving Security of SPN-Type Block Cipher Against
Fault Attack** 115
Gitika Maity, Sunanda Jana, Moumita Mantri and Jaydeb Bhaumik

**High-Capacity Reversible Data Hiding Scheme Using Dual Color
Image Through (7, 4) Hamming Code** 127
Ananya Banerjee and Biswapati Jana

**A Study on the Effect of a Rectangular Slot on Miniaturization of
Microstrip Patch Antenna** 141
Sunandan Bhunia and Avisankar Roy

**FPGA Implementation of OLS (32, 16) Code and
OLS (36, 20) Code** 151
Arghyadeep Sarkar, Jagannath Samanta, Amartya Barman
and Jaydeb Bhaumik

**Comparative Study of Wavelets for Image Compression with
Embedded Zerotree Algorithm** 163
Vivek Kumar and Govind Murmu

**Design of Compact Wideband Folded Substrate-Integrated
Waveguide Band-Pass Filter for X-band Applications** 173
Nitin Muchhal, Abhay Kumar, Arnab Chakrabarty
and Shweta Srivastava

DCT-Based Gray Image Watermarking Scheme 181
Supriyo De, Jaydeb Bhaumik, Puja Dhar and Koushik Roy

**Five-Input Majority Gate Design with Single Electron
Nano-Device** 191
Arpita Ghosh and Subir Kumar Sarkar

Post-layout Power Supply Noise Suppression and Performance Analysis of Multi-core Processor Using 90 nm Process Technology 199
Partha Mitra and Jaydeb Bhaumik

Enhanced Performance of GaN/InGaN Multiple Quantum Well LEDs by Shallow First Well and Stepped Electron-Blocking Layer 207
Mainak Saha and Abhijit Biswas

A μ-Controller-Based Biomedical Device to Measure EMG Strength from Human Muscle 217
Arindam Chatterjee

Design and Implementation of a DCM Flyback Converter with Self-biased and Over-Current Protection Circuit 229
R. Rashmi and M. D. Uplane

Performance Improvement of Light-Emitting Diodes with W-Shaped InGaN/GaN Multiple Quantum Wells 241
Himanshu Karan and Abhijit Biswas

Behavioral Modeling of Differential Inductive Seismic Sensor and Implementation of Its Readout Circuit 253
Abhishek Kumar Gond, Rajni Gupta, Samik Basu, Soumya Pandit
and Soma Barman

Application of PSO Variants for Optimal Design of Two-Stage CMOS Op-amp with Robust Bias Circuit 263
Bishnu Prasad De, K. B. Maji, Dibyendu Chowdhury, R. Kar,
D. Mandal and S. P. Ghoshal

Representation and Exploring the Semantic Organization of Bangla Word in the Mental Lexicon: Evidence from Cross-Modal Priming Experiments and Vector Space Model 273
Rakesh Dutta, Biswapati Jana and Mukta Majumder

Solving a Solid Transportation Problems Through Fuzzy Ranking 283
Sharmistha Halder(Jana), Barun Das, Goutam Panigrahi
and Manoranjan Maiti

Optimal Design of Low-Noise Three-Stage Op-amp Using PSO Algorithm 293
K. B. Maji, B. P. De, R. Kar, S. P. Ghoshal and D. Mandal

Optimal Design of Low-Voltage, Two-Stage CMOS Op-amp Using Evolutionary Techniques 303
Bishnu Prasad De, K. B. Maji, Banibrata Bag, Sayan Tripathi, R. Kar,
D. Mandal and S. P. Ghoshal

Message from the Volume Editors

It is a great pleasure for us to organize the first International Conference on *Communication, Devices, and Computing* (ICCDC 2017) held during November 2–3, 2017, at the Haldia Institute of Technology, Purba Medinipur, West Bengal, India. Our main goal is to provide an opportunity to the participants to learn about contemporary research in the area of *Communication, Devices, and Computing* and to exchange ideas among themselves and with experts present in the conference as invited speakers. It is our sincere hope that the conference will help the participants in their research and training. Also will open new avenues of work for those who are either starting their research or looking for extending their area of research in *Communication, Devices, and Computing*.

After an initial call for papers, 62 papers were submitted for presentation at the conference. All submitted papers were sent to external referees, and after refereeing, 29 papers were recommended for publication in the conference proceedings which will be published by Springer in its Lecture Notes on Electrical Engineering (LNEE) series.

We are grateful to the speakers, participants, referees, organizers, sponsors, and funding agencies (DRDO, ISRO) for their support and help, without which it would have been impossible to organize this conference. We express our gratitude to the organizing committee members who work behind the scene tirelessly to make this conference successful.

About the Editors

Jaydeb Bhaumik is a Professor in the Department of Electronics and Communication Engineering, Haldia Institute of Technology, Haldia, West Bengal, India. He received his Ph.D. from IIT Kharagpur in 2010. He has 2 books, 19 journal papers, 25 conference papers, and 1 technical report to his credit. He has 10 years of teaching and over 4 years of research experience. His areas of interest include VLSI architectures for cryptographic algorithms and error correcting codes, security issues in image processing, lightweight block ciphers, and cellular automata.

Indrajit Chakrabarti, Ph.D., is a Professor in the Department of Electronics and Electrical Communication Engineering, Indian Institute of Technology Kharagpur, India. He holds 2 patents and has published 1 book, 33 journal papers, and 73 conference papers. His primary research interest is in VLSI design.

Dr. Bishnu Prasad De is an Assistant Professor in the Department of Electronics and Communication Engineering, Haldia Institute of Technology, West Bengal, India. He has 7 years of teaching and 2 years of research experience. His areas of interest include VLSI circuits and systems, analog electronics, electronic design automation, and soft computing. He has published several papers in journals of national and international repute.

Mr. Banibrata Bag is an Assistant Professor in the Department of Electronics and Communication Engineering, Haldia Institute of Technology, West Bengal, India. He has 6 years of teaching and 4 years of industry experience. His main area of research is optical wireless communications and networks. He has published one book, fivejournal papers, and nine conference papers.

Mr. Surajit Mukherjee is an Assistant Professor in the Department of Electronics and Communication Engineering, Haldia Institute of Technology, West Bengal, India. He has 7 years of teaching experience and has published six journal papers and four conference papers. His areas of research include antenna design, microwave filter design, and frequency selective surfaces.

Binary Error Correcting Code for DNA Databank

Jagannath Samanta, Jaydeb Bhaumik, Soma Barman and Raj Kumar Maity

Abstract Deoxyribonucleic acid (DNA) is the storage media of heredity information of all living species. These DNA molecules store the digital information that comprises the genetic blueprint of living organisms. The DNA databank can be employed in the analysis of genetic diseases, fingerprinting for criminology or genetic genealogy. There is the possibility of soft errors in these DNA databases during the long-term storage purpose. In this chapter, a binary error correcting code (ECC) is used to improve the reliability of stored DNA bases. An encoding and decoding algorithm is proposed for any arbitrary length of DNA sequences. A single error correcting code (280, 256) codec has been designed and implemented. Proposed designs are simulated and synthesized using both FPGA and ASIC platforms.

Keywords Error correcting code · Single error correcting code
Deoxyribonucleic acid · FPGA and ASIC

1 Introduction

Deoxyribonucleic acid (DNA) was discovered by Swiss physician Friedrich Miescher in 1868. DNA is a molecule which encodes an organism's genetic blueprint; i.e., it stores all of the information needed to build and maintain a living organism which is realized through a process discussed by the "Central dogma of molecular biology" [1]. This digital information is generated and reliably stored in coding

J. Samanta (✉) · J. Bhaumik · R. K. Maity
Haldia Institute of Technology, Haldia, India
e-mail: jagannath19060@gmail.com

J. Bhaumik
e-mail: bhaumik.jaydeb@gmail.com

R. K. Maity
e-mail: hitece.raj@gmail.com

S. Barman
Institute of Radio Physics and Electronics, University of Calcutta, Kolkata, India
e-mail: barmanmandal@gmail.com

© Springer Nature Singapore Pte Ltd. 2017
J. Bhaumik et al. (eds.), *Communication, Devices, and Computing*, Lecture Notes in Electrical Engineering 470, https://doi.org/10.1007/978-981-10-8585-7_1

regions of DNA sequences throughout the billions of years. These DNA sequences are remarkably preserved despite striking differences in the body plans of different animals. In DNA, biomolecular base pairs are existed tightly together. The four DNA bases are adenine (a), thymine (t), guanine (g), and cytosine (c) [2]. These bases are categorized into two types of base pairs, i.e., ag and ct [2]. Therefore, it is logical to represent these DNA base pairs in terms of a binary linear code without loss of its generality [2].

The DNA is efficiently organized into long structures called chromosomes. These set of chromosomes in a cell create its genome. The human genome has around three billion base pairs of DNA arranged into forty-six chromosomes [2, 3]. The information carried by the region of DNA is called gene. The human beings have nearly 25,000 genes. Each gene of human DNA has an average length of 3000 bases. The DNA research has shown its tremendous scientific and medical progress during last few years. A DNA databank or DNA databases are generally employed to examine the genetic diseases, genetic fingerprinting, forensic sciences, etc. Recently, the concept of error correcting code (ECC) has been employed in DNA sequencing for detection and diagnosis of harmful diseases [4, 5].

Figure 1 shows the Gatlin's communication theory model for the genetic system. Gatlin represented this genetic system as an error control coding models. It described that extra bases in DNA may be employed for error detection and correction purposes [6]. The encoded DNA moves through a channel (transcription and translation) which transfers all the mechanics for amino acid protein sequence production [6]. Finally, generated protein is treated as received sequence.

Several other classical and quantum biological channel models like May model, Markovian classical model, Markovian quantum model, hybrid quantum/classical model, multilevel symmetric channel model, and Kimura model are discussed in literature [1, 2, 7].

Figure 2 shows the framework for DNA sequence with channel coding used in storage mediums. The DNA of different spices are stored in DNA databank in the form of chain of nucleotide bases a, t, g, and c which are mapped into binary equivalent. Then, code words are generated by employing conventional channel encoding process. These code words may be affected by soft errors during the storing in different storage mediums. The original code words are extracted from received code words through channel decoding block. Then, the corrected DNA bases are obtained using reverse binary mapping process which is shown in Fig. 2 (Tables 1 and 2).

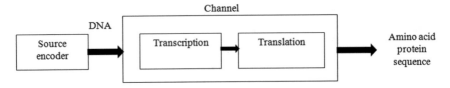

Fig. 1 Gatlin's communication theory model for the genetic system

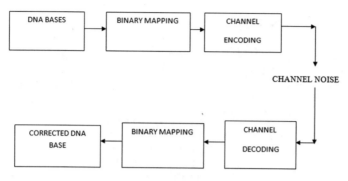

Fig. 2 Framework for DNA sequence with channel coding

Table 1 Sample SEC (280, 256) encoder output

DNA base	Binary representation	Parity
aggc tcac tata aata gcag ccac ctct ccct ggca gaca ggga cccg cagc tcag ctac agca caga tcag cacc atga agct tctc acgg gcct ggtt ttct gctc cttg gtcc tgag tgtc agca	00101011 01110011 01000100 00000100 10110010 11110011 11011101 11111101 10101100 10001100 10101000 11111110 11001011 01110010 11010011 00101100 11001000 01110010 11001111 00011000 00101101 01110111 00111010 10111101 10100101 01011101 10110111 11010110 10011111 01100010 01100111 00101100	**11001101** **11001110** **11111101**
tgag gtct atgt ccag agaa gctg agat atgg cata taat aggc atct aata aatg ctta agag gtgg aatt tgtt gaaa, ca **atgc atgc atgc atgc atgc atgc atgc atgc atgc atgc at**	01100010 10011101 00011001 11110010 00100000 10110110 00100001 00011010 11000100 01000001 00101011 00011101 00000100 00000110 11010100 00100010 00000101 01100101 10000000 1100 00011011 00011011 00011011 00011011 00011011 00011011 00011011 00011011 00011011 00011011 00011011 0001	**10010010** **00000111** **01000111**

In this chapter, we have proposed an encoding and decoding steps for arbitrary DNA sequence. Reviriego et al. SEC (280, 256) code is used for channel encoding and decoding of DNA databases [8].

The rest of chapter is organized as follows. Different types of DNA databases and its importance are discussed in Sect. 2. Proposed Encoding and decoding algorithm is described in Sect. 4. In Table 3, FPGA and ASIC synthesis results are presented in Table 4. Finally, the chapter is concluded in Sect. 5.

Table 2 Sample SEC (280, 256) decoder output

Received code word	Decoded sequence in binary	DNA base
00101011 01110011 01000100 00000100 10110010 11110011 11011101 11111101 10101100 10001100 10101000 11111110 11001011 01110010 11010011 00101100 11001000 01110010 11001111 00011000 00101101 01110111 00111010 10111101 1010010101011101 10110111 11010110 10011111 01100010 01100111 00101100 11001101 11001110 11111101	00101011 01110011 01000100 00000100 10110010 11110011 11011101 11111101 10101100 10001100 10101000 11111110 11001011 01110010 11010011 00101100 11001000 01110010 11001111 00011000 00101101 01110111 00111010 10111101 10100101 01011101 10110111 11010110 10011111 01100010 01100111 00101100	aggc tcac tata aata gcag ccac ctct ccct ggca gaca ggga cccg cagc tcag ctac agca caga tcag cacc atga agct tctc acgg gcct ggtt ttct gctc cttg gtcc tgag tgtc agca
01100010 10011101 00011001 011110010 00100000 10110110 00100001 00011010 11000100 01000001 00101011 00011101 00000100 00000110 11010100 00100010 00000101 01100101 10000000 1100000 11011 00011011 00011011 00011011 00011011 00011011 00011011 00011011 00011011 00011011 00011011 0001 10010010 00000111 01000111	01100010 10011101 00011001 11110010 00100000 10110110 00100001 00011010 11000100 01000001 00101011 00011101 00000100 00000110 11010100 00100010 00000101 0110010 110000000 1100 00011011 00011011 00011011 00011011 00011011 00011011 00011011 00011011 00011011 00011011 00011011 0001	tgag gtct atgt ccag agaa gctg agat atgg cata taat aggc atct aata aatg ctta agag gtgg aatt tgtt gaaa ca atgc atgc atgc atgc atgc atgc atgc atgc atgc atgc atgc at

Table 3 FPGA synthesis of Reviriego et al. SEC (280, 256) codec

Specification	Encoder	Decoder
Number of slices	100	252
Number of 4 input LUTs	177	446
Number of bonded IOBs	280	536
Maximum combinational path delay (ns)	6.493	7.42

Table 4 ASIC synthesis of Reviriego et al. SEC (280, 256) codec

Specification	Encoder	Decoder
Area (in terms of gates)	929	1849
Clock frequency (MHz)	215.6	204.7

2 DNA Databank or DNA Database

DNA can be employed to recognize criminals with greater accuracy when the biological evidence exists [9]. The DNA databases help to get an clear idea of suspects, mistakenly accused or convicted for a particular crime [10]. The National Center for Biotechnology Information (NCBI) stores the DNA sequence of living beings and encourages the scientific community to access these databases for their usage in biomedical and genomic research [11].

2.1 Types of DNA Databases

National DNA Databank: National DNA databank is retained by the government for storing DNA profiles of its people. These databases are largest ones, and most of people use this database. Tandemly repeated DNA sequences are extensive throughout the human genome. These sequences are important genetic markers for mapping studies, disease diagnosis, and human identity testing. Short tandem repeats (STRs) hold the repeat units of 2–6 base pair and can be quickly amplified with the polymerase chain reaction (PCR). STRs have become favorable in forensic laboratories due to its low amounts of DNA sequence requirement. These DNA profiles are generally employed for forensic purposes which include searching and matching of potential criminal suspects [12].

Forensic DNA Databank: Forensic database is used to produce the matches between the crime scene biomarkers and suspected individual [13]. A centralized database is maintained for individuals storing DNA profile that enables searching and detecting of DNA samples collected from a crime scene against each stored profiles. The forensic databases produce evidence to support criminal investigations and also identify potential suspects. Most of the national DNA databases are employed for forensic purposes [14]. The largest and most popular forensic database in the world is the UK national DNA database (NDNAD) [15]. In NDNAD, only patterns of short tandem repeats are stored instead of individual's full genomic sequence [15].

Medical DNA Databank: The medical DNA database contains medically relevant genetic variations. This database collects individual's DNA that may reflect their medical and lifestyle details. Researchers may discover the interactions between the genetic environment and occurrence of certain diseases through recording of DNA profiles. Finally, they may find some new drugs or effective treatments in controlling critical diseases.

Importance of DNA databank has been described in the following section.

3 Importance of DNA Databank

DNA analysis may be used to identify specific human genes which is responsible for triggering major diseases, and based on this, new drugs come into the market. These identified genes and their successive analysis in terms of therapeutic treatment have finally influenced the medical science research.

Disease diagnosis and treatment: Due to vast research in DNA sequence, there is a possibility of ease diagnosis of diseases in molecular level. Now, DNA research has essentially led to promising drugs, treatments, and therapies for dreadful diseases. Researchers also find to formulate the brand new drugs to treat these diseases. Many countries accumulate newborn blood samples to identify diseases related to genetic abnormalities [16].

Paternity and legal impact: Paternity cases have a broad impact on families and children around the world. The paternity of a child can be identified through the assessment of DNA. This has a significant effect on the child's future life [16, 17].

Forensics science: The innocence or guilt of a person can be investigated through forensics science. This technique is very important for victims' identification, particularly where the victim's external features are unidentifiable to the family members or friends [16]. Criminal or anti-socialist can also be detected using DNA fingerprinting [16]. The Interpol DNA database is used in criminal investigations. In all, DNA technology is progressively essential to ensure accuracy and fairness in the criminal justice system [10].

Twins: DNA is highly discriminatory which can make it a powerful tool to identify the individuals. Each person's DNA is unique, even if they are twins. Now, twins have incredibly small mutations which can be detected. Tiny differences between identical twins can now be detected by next-generation sequencing. The scientists examined the genes of 10 pairs of identical twins, including 9 pairs in which one twin showed signs of dementia or Parkinson's disease and the other did not matched [18].

Agriculture: The effect of DNA in agriculture is a very important because it facilitates the breeding of plants or animals that have a better resistance to diseases. It also allows farmers to produce more nutritious products [16].

DNA Computing: DNA computing is a promising technology for current and future applications. Research is going on DNA computer which can be placed in human body to diagnosis disease cells and kill them. A computer is built using DNA circuits rather than silicon circuits [16].

In the twentieth century, the discovery of DNA has been a significant impact on society. Nowadays, it will continue to reform in medicine, agriculture, forensics, and many other important emerging fields. There is a possibility of soft errors in these DNA databases during the long-term storage purposes. These errors can reduce the reliability of these DNA databases. The error correcting codes (ECCs) are the suitable techniques which can reduce the soft errors in storage systems containing DNA databases.

In this chapter, binary error correcting codes (ECCs) are used to improve the reliability of stored DNA bases. These ECC encoder and decoder circuits require extra circuitry due to additional parity bits. Design of ECC encoders and decoders require extra circuitry due to additional parity bits. As a result, the hardware complexity, power requirements, and delay of ECC circuit will also be affected. So the main concern of researchers is to reduce the number of required parity bits in ECC encoder and decoder circuits.

Single error correcting (SEC) codes can correct single bit error per data block. The SEC codes are comparatively simpler than single error correcting-double error detecting (SEC-DED) code. Implementation of binary ECC is presented in the following section.

4 Binary ECC Code for DNA Sequences

One most important SEC code is Reviriego et al. code with weight two [8]. In general, DNA sequence length is in the order of thousand and its length is not fixed. Here, we have taken the Homo sapiens DNA sequences with NCBI reference sequence number NM-030754.4 with 594 base pair (Homo sapiens serum amyloid A2 (SAA2), transcript variant 1, mRNA) as a sample database [11] which is shown in Fig. 3. For storing this DNA sequence, Hamming SEC codec is used. The DNA bases are represented as $a = 00$, $t = 01$, $g = 10$, and $c = 11$. For space constraints, the first part 1–56 base pair (bp) and last part 582–594 bp are displayed in this chapter. In the following subsection, we have proposed an encoding and decoding algorithm for any arbitrary DNA sequence length.

4.1 Proposed SEC Encoding Steps

To encode given DNA base sequences into (n, k) block codes, the length of DNA base sequence must be an integer multiple of k. However, this is not possible because the

---

```
  1 aggctcacta taaatagcag ccacctctcc ctggcagaca gggacccgca gctcagctac
 61 agcacagatc agcaccatga agcttctcac gggcctggtt ttctgctcct tggtcctgag
121 tgtcagcagc cgaagcttct tttcgttcct tggcgaggct tttgatgggg ctcgggacat
181 gtggagagcc tactctgaca tgagagaagc caattacatc ggctcagaca aatacttcca
241 tgctcggggg aactatgatg ctgccaaaag gggacctggg ggtgcctggg ctgcagaagt
301 gatcagcaat gccagagaga atatccagag actcacaggc cgtggtgcgg aggactcgct
361 ggccgatcag gctgccaata aatggggcag gagtggcaga gaccccaatc acttccgacc
421 tgctggcctg cctgagaaat actgagcttc ctcttcactc tgctctcagg agacctggct
481 atgaggccct cggggcaggg atacaaagtt agtgaggtct atgtccagag aagctgagat
541 atggcatata ataggcatct aataaatgct taagaggtgg aatttgttga aaca
```

Fig. 3 DNA samples from NCBI

length always attains random values and chances of length being integer multiple of k are very less. To cope with this situation, an encoding algorithm is being proposed where some extra bases are added deliberately such that the length of DNA base sequence becomes an integer multiple of k. Here, "atgcatgc" is chosen as redundant base sequence. The redundant bases to be added are same for every sequence in order to detect them while decoding. This addition of redundant base process will be repeated until the sequence length becomes integer multiple of k.

Steps of proposed SEC encoding
1. *Let, l be the length of a given DNA base sequence*
2. *Number of bits after binary mapping = 2l (as two bits are used for each base)*
3. *Maximum number of code word, $N = \lceil (2l/k) \rceil$*
4. *Number of valid bits in the last code word $R = 2l \bmod k$*
5. *Number of redundant bits to be added = $k - R$*
6. *Length of DNA base sequence to be added = $(k - R)/2$*
7. **for** *($i = 1$ to $N - 1$)* **do**
   | *Parity bits are generated for each block of k bit and appended with the sequence to generate*
   | *the code word.*
end
8. **for** *($i = N$)* **do**
   | *$(k - R)$ bit are appended with the rest of the sequence, and then, parity bits are generated.*
   | *Then the last code word is obtained.*
end

4.2 *Proposed SEC Decoding Steps*

Decoding method is almost similar to the conventional decoding method being used so far except the extra step for checking and detecting the redundant bits (if any) in sequence. If decoded sequence ends with "atgcatgc," then we can say that these are redundant bases and must be removed in order to get the original sequence.

Steps of proposed SEC-based decoding
1. *Length of received code word in bit = $N * n$*
2. *Length of sequence = $N * k$*
3. *Redundant bits = $k - R$*
4. *Length of original sequence = $N * k - (k - R)$*
5. **for** *($i = 1$ to $N - 1$)* **do**
   | *Perform like the conventional SEC code*
end
6. **for** *($i = N$)* **do**
   | *Perform like conventional decoding operation.*
   | *If "0001101100011011" pattern is found then discard rest of the sequence, to obtain the*
   | *correct DNA sequence*
end

Reviriego et al. SEC (280, 256) code of weight-two scheme is designed for any large length DNA sequence by employing proposed encoding and decoding scheme.

4.3 Reviriego et al. SEC (280, 256) Code for Any Arbitrary DNA Sequence

Reviriego et al. SEC (280, 256) code [8] has been implemented in this subsection. A code word of 280 bit is generated by appending 24-bit parity with 256 bit of DNA sequence. Here, same DNA sequence has been considered like in Hamming SEC (12, 8) code.

The H-matrix of SEC (280, 256) code is given in the Fig. 4. A total of 24 parity bits are obtained from this H-matrix. Width of all the columns of data bits of this H-matrix is always two. The total numbers of 1's in each row is shown in 3rd part, last column of H-matrix of SEC (280, 256) code in Fig. 4.

A gate level design of SEC (280, 256) encoder is shown in Fig. 5. Input bits $i_0, i_1, \ldots, i_{254}, i_{255}$ and parity bits $p_1, p_2, \ldots, p_{23}, p_{24}$ are represented in Fig. 5. The SEC (280, 256) encoder requires a total of 512 numbers of XOR2 gates.

This SEC (280, 256) encoder will generate the parity bits, and sample output is presented in Table 1. Here, first and second rows of Table 1 show the DNA sequence of 1–128 bp and 513–594 bp, respectively. DNA sequences in bold in second row and first column are the redundant bits.

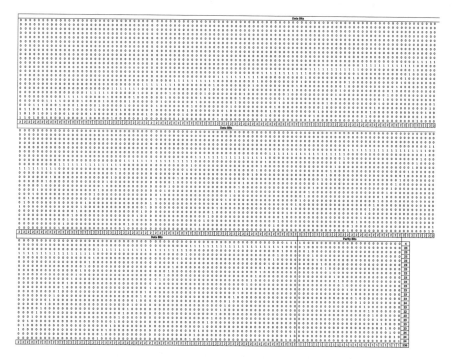

Fig. 4 H-matrix of SEC (280, 256) code

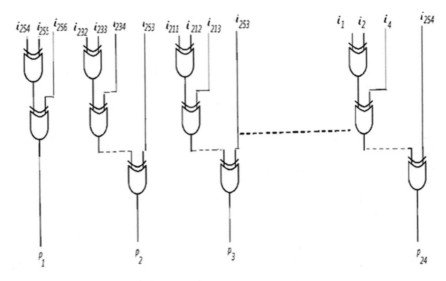

Fig. 5 Gate level design of SEC (280, 256) encoder

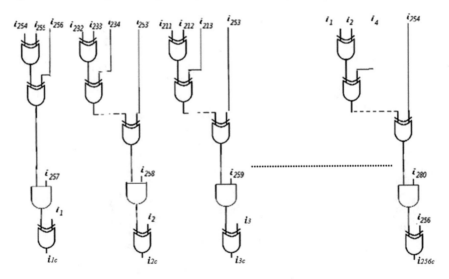

Fig. 6 Gate level design of SEC (280, 256) decoder

Similarly, a gate level design of SEC (280, 256) decoder is shown in Fig. 6. SEC (280, 256) decoder circuit is more complex compared to encoder circuit. The SEC (280, 256) decoder requires total numbers of 792 XOR2 gates and 256 AND2 gates. Decoder detects and corrects single bit error. Sample decoder output is presented in Table 2.

4.3.1 Implementation of Reviriego et al. SEC (280, 256) Codec

FPGA-based synthesis result of Reviriego et al. SEC (280, 256) codec is shown in Table 3. Area and path delay information of SEC (280, 256) codec is observed in Table 3.
ASIC-based synthesis result of Reviriego et al. SEC (280, 256) codec is shown in Table 4. Area in terms of logic gates and slack time is also observed in Table 4.

5 Conclusion

In this chapter, we have proposed an encoding and decoding procedure using Reviriego et al. SEC code to enhance the reliability of stored DNA sequences. VLSI implementation results of SEC (280, 256) codec are presented. Synthetic DNA sequence is protected using binary error correcting codes. Soft errors in DNA databank can easily be identified by employing these ECCs because one base error in DNA sequence corresponds to two-bit positions. Binary ECC increases the robustness and reliability of these DNA storage mediums. Application of error correcting codes for DNA sequence can be beneficial for forensic sciences and medical sciences. This approach provides a cost-effective way of DNA storage.

References

1. E.E. May, Comparative analysis of information based models for initiating protein translation in Escherichia coli K-12, M.S. thesis, NCSU, Dec 1998
2. E. May, M. Vouk, D. Nitzer, D. Rosnick, An error-correcting code framework for genetic sequence analysis. J. Franklin Inst. **34**, 89–109 (2004)
3. L. Liebovitch, Y. Tao, A. Todorov, L. Levine, Is there an error correcting code in the base sequence in DNA? Biophysical J. **71**(3), 1539–1544 (1996)
4. M. Blawat, K. Gaedke, I. Huetter, X.M. Chen, B. Turczyk, S. Inverso, B.W. Pruitt, G.M. Church, Forward error correction for DNA data storage. Procedia Comput. Sci. **80**, 1011–1022 (2016)
5. L.C.B. Faria, A.S.L. Rocha, J.H. Kleinschmidt, R. Palazzo, M.C. Silva-Filho, DNA sequences generated by BCH codes over GF(4). Electron. lett. **46**(3), 203–204 (2010)
6. L.L. Gatlin, *Information Theory and the Living System* (Columbia University Press, New York, NY, 1972)
7. I.B. Djordjevic, Classical and quantum error-correction coding in genetics, *Quantum Biological Information Theory*, pp. 237–269 (2016)
8. P. Reviriego, S. Pontarelli, J.A. Maestro, M. Ottavi, A method to construct low delay single error correction codes for protecting data bits only. IEEE Trans. Comput. Aided Des. Integ. Circ. Syst. **32**(3), 479–483 (2013)
9. J.L. Doleac, The effects of DNA databases on crime. Am. Econ. J. Appl. Econ. Life Sci. Soc. Policy **9**(1), 165–201 (2017)
10. Advancing justice through DNA technology using DNA solve crimes, U.S. Department of Justice Archives. https://www.justice.gov/archives/ag
11. https://www.ncbi.nlm.nih.gov/nuccore/315075328/

12. I. Murnaghan, The importance of DNA, explore DNA (2017). http://www.exploredna.co.uk/the-importance-dna.html
13. F. Santos, H. Machado, S. Silva, Forensic DNA databases in European countries: is size linked to performance? Life Sci. Soc. Policy **9**(1), 4513018 (2013)
14. J.M. Butler, *Advanced Topics in Forensic DNA Typing: Methodology* (Academic Press, 2011)
15. H. Wallace, The UK national DNA database: balancing crime detection, human rights and privacy, EMBO Report, pp. S26–S30 (2006)
16. http://www.exploredna.co.uk/the-importance-dna.html
17. J.L. Doleac, The effects of DNA databases on crime. Am. Econ. J. Appl. Econ. **9**(1), 165–201. https://doi.org/10.1257/app.20150043. ISSN 1945-7782
18. A. Oconnor, The claim: identical twins have identical DNA, The New York Times (2008). http://www.nytimes.com/2008/03/11/health/11real.html

Proactive and Reactive DF Relaying for Energy Harvesting Underlay CR Network

Mousam Chatterjee, Subhra Shankha Bhattacherjee and Chanchal Kumar De

Abstract Outage performance analysis of energy harvesting underlay CR network using proactive and reactive DF relay selection schemes in presence of non-identical Rayleigh fading channel has been studied in this paper. Time switching relay (TSR) protocol is used for the combined task of energy harvesting and signal transmission at relays. The relay can harvest energy from source signal as well as primary transmitter interference. The outage probability has been evaluated at the secondary destination. Selection combining (SC) is used at relays to select the best among them. The effects of number of relay antennas (L) and number of relays (K) on the outage performance for both proactive and reactive DF relaying have also been studied. A trade-off between number of relay and number of antenna at each relay is studied. It is also shown that the reactive DF scheme may generate different outage probability for different values of threshold SNR (μ_{th}) for successful decoding at the relays, which may be sometimes better or worse than the proactive DF scheme.

Keywords Proactive DF · Reactive DF · Energy harvesting · Selection combining · Outage probability

1 Introduction

Communication spectrum is a limited asset, and its demand is steadily growing, both because billions of new users are joining the Internet and new apps are using more bandwidth. In an underlay environment, there are two kinds of users—first a licensed user who is called a primary user and second a non-licensed user who

M. Chatterjee
B. P. Poddar Institute of Management and Technology, Kolkata, India
e-mail: mousam.chatterjee@gmail.com

S. S. Bhattacherjee
Dr. B. C. Roy Engineering College, Durgapur, India
e-mail: subhra54@gmail.com

C. K. De (✉)
ECE Department, Haldia Institute of Technology, Haldia, India
e-mail: wrt2chanchal@rediffmail.com

© Springer Nature Singapore Pte Ltd. 2017
J. Bhaumik et al. (eds.), *Communication, Devices, and Computing*, Lecture Notes in Electrical Engineering 470, https://doi.org/10.1007/978-981-10-8585-7_2

is called a secondary user. The secondary users can transmit signals using the primary spectrum band by maintaining an interference constraint for the primary user. Cognitive radio (CR) is considered to be a promising technology for implementing opportunistic spectrum access [1]. Thus performance analysis of underlay CR network has gained research interest in our work. Energy conscious communication has received considerable attention in recent years [2, 3]. Green communication techniques are developed to judiciously use the available resources [4, 5]. Authors in [6] have studied different relay selection strategies and evaluated the outage probabilities. Closed-form expression of DF relay is derived over Rayleigh fading channels in [7]. In [8], authors have investigated a dual-hop spectrum sharing protocol based on DF relay, where it is assumed that the cognitive relays are equipped with multiple antennas. Authors in [9] have compared the outage probabilities of proactive and reactive DF relay selection schemes under Rayleigh fading channel. Exact outage probabilities of proactive DF relay selection with MRC receivers have been derived in [10]. Asymptotic outage probabilities have been calculated for proactive and reactive schemes in presence of multiple relays in [11]. Performance analysis of routing protocols in mobile ad hoc networks or MANETS has been done in [12]. Power allocation in RF energy harvesting DF relay has been studied in [13]. Outage and throughput calculations for RF energy harvesting-based two-way DF relay is evaluated in [14]. Analytical expressions for outage probability for proactive and reactive relay selection schemes in presence of multiple primary users and multi-antenna-based relays have been derived in [15]. In the above-mentioned works, a typical scenario has not been considered where proactive and reactive relay selection schemes coexist with energy harvesting at the relays.

In this paper, an underlay paradigm has been considered, which consists of one secondary user (SS) source, SR_K multi-antenna-based cognitive relays, one secondary destination (SD), one primary transmitter (PU_{Tx}), and one primary receiver (PU_{Rx}). The relays are equipped with energy harvesting circuits and rechargeable battery. The relays can harvest energy from source signal and interference from primary transmitter and use this energy to forward the received signal to destination which is in contrast to [15] where energy harvesting has not been considered. It is also assumed that the source and the destination terminals are equipped with single antenna, while multiple antennas are used at the relays. In this work, proactive and reactive relay selection schemes are used. Outage probability has been considered as the performance metric, and the performance of the network using proactive and reactive relay selection schemes is analyzed with changing parameters. More precisely, our contributions in this paper are

1. Evaluation of outage performance of a underlay energy harvesting CR network with the help of a multi-antenna-based proactive and reactive relays in presence of non-identical Rayleigh fading channel.
2. Analyzing the impact of increasing number of antennas (L) and number of relays (K) on outage performance for both reactive and proactive schemes.
3. A trade-off between number of relays and number of antennas at each relay has been shown.

4. The effect of varying values of μ_{th} on reactive outage probability and comparing it with proactive outage probability.

This paper is organized as follows—Sect. 2: discussion on the system model, Sect. 3: description of the simulation model, Sect. 4: analysis of the results, and finally conclusions are drawn in Sect. 5.

2 System Architecture

2.1 Network Model

The network configuration shown in Fig. 1 is considered as the system model. It consists one secondary source SS, one secondary destination SD, multiple secondary relays SR_k (where $k = 1, 2, \ldots, K$) equipped with l number of antennas where $l = 1, 2, \ldots, L$, one primary transmitter PU_{Tx}, and one primary receiver $PU - Rx$. The considered CR network operates in underlay mode which implies that the SS can send signal to SD by maintaining an interference constraint I_P for the primary receiver. No direct link has been considered between secondary source and secondary destination due to severe shadowing. The primary transmitter interference is considered on the secondary relays. The relays are rigged with energy harvesting circuits which are capable of harvesting energy from ambiance RF signal. This harvested energy is stored in rechargeable battery or super-capacitor and further used in signal transmission. The relays harvest energy from interference signal due to PU_{Tx} and incoming source signal. Multiple antennas are used at relays to increase SNR diversity from SS to SR. The channel coefficients follow Rayleigh distribution with zero mean and unit variance, and all the noise terms are zero mean additive white Gaussian noise (AWGN) with variance N_o. The channel link coefficients are

Fig. 1 System model for underlay CR network with multi-antenna-based relay

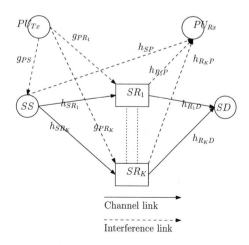

Channel link

Interference link

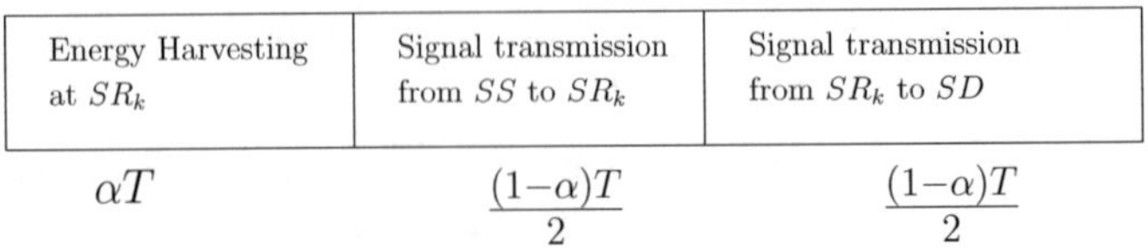

Energy Harvesting at SR_k	Signal transmission from SS to SR_k	Signal transmission from SR_k to SD
αT	$\dfrac{(1-\alpha)T}{2}$	$\dfrac{(1-\alpha)T}{2}$

Fig. 2 Time switching-based relaying protocol

described as: $h_{SR_{k,l}}$ for SS to l_{th} antenna of k_{th} relay $R_{k,l}$; $h_{SR_k} = \max\limits_{l=1,2,...,L} \{h_{SR_{K,l}}\}$, h_{R_kD} for R_k to SD, h_{SP} is for SS to PU_{Rx}, h_{R_kP} is for SR to PU_{Rx}, and g_{PR_k} is for SR_k to PU_{Tx}. Proactive and reactive DF relay selection schemes are used for selecting the best relay to forward the source signal to the destination.

2.1.1 Power Allocation at Source and Relays

The power allocated at the secondary source is such that it maintains an interference constraint I_p for the primary receiver, Thus the power at source is given by

$$P_{SS} = \frac{I_p}{|h_{SP}|^2} \tag{1}$$

As shown in Fig. 2, the relays use TSR protocol [13] to jointly complete the task of signal transmission and energy harvesting. The relays can harvest energy from source signal and from PU_{Tx} interference signal for a time period of αT. In second time slot, i.e., $(1 - \alpha)T$, this energy is used for signal transmission purpose. For $\frac{(1-\alpha)}{2}T$ the source sends signal to relay and for next half of time slot, i.e., $\frac{(1-\alpha)}{2}T$, the relay sends signal to destination. The harvested energy at kth relay due to source signal is given by [16]

$$E_{R_k}^{Source} = \frac{\eta P_{SS}|h_{SR_k}|^2 \alpha T}{d_1{}^n} \tag{2}$$

and energy harvested due to interference signal of PU_{Tx} on SR is given by [16]

$$E_{R_k}^{Inf} = \frac{\eta P_p|g_{PR_k}|^2 \alpha T}{d_2{}^n} \tag{3}$$

where P_p is the primary user operating power, P_{SS} is allocated source power, d_1 and d_2 are the distances between secondary source to relay and relay to primary transmitter, respectively, n is the path loss exponent and $0 < \alpha < 1$ and η is the energy conversion efficiency $0 < \eta < 1$.

Thus, the total energy harvested at each relay is

$$E_{R_k} = E_{R_k}^{Inf} + E_{R_k}^{Source} \qquad (4)$$

The maximum power allocated at each relay due to harvested energy is [13]

$$P_{R_k}^{Eh} = \frac{E_{R_k}}{(1-\alpha)\frac{T}{2}} \qquad (5)$$

The power allocated at kth relay SR_k by maintaining interference constraint I_p for the primary receiver is

$$P_{R_k}^{Inf} = \frac{I_p}{\left|h_{R_kP}\right|^2} \qquad (6)$$

where $k = 1, 2, \ldots, K$.

Therefore, the effective power allocated at each secondary relay SR is given by

$$P_{R_k} = \min_{k=1,2,\ldots,K} \{P_{R_k}^{Eh}, P_{R_k}^{Inf}\} \qquad (7)$$

2.2 Relay Selection Schemes

In this section, we will define the proactive and reactive DF relay selection schemes and evaluate their outage probabilities.

2.2.1 Proactive DF Relay Selection

According to proactive DF scheme [15], the best SR is chosen by maximizing the minimum SNR between $SS \rightarrow SR_k$ and $SR_k \rightarrow SD$ where $k = 1, 2, \ldots, K$. Thus, the end-to-end SNR is given by

$$\gamma^{Pro} = \max_{k \in 1,2,\ldots,K} \{min\left(\gamma_{SR_k}, \gamma_{R_kD}\right)\} \qquad (8)$$

where $\gamma_{SR_k} = \frac{P_{SS}}{N_o}\left|h_{SR_k}\right|^2 = \frac{I_p}{N_o}\frac{|h_{SR_k}|^2}{|h_{SP}|^2}$ and $\gamma_{R_kD} = \frac{P_{R_k}}{N_o}\left|h_{R_kD}\right|^2$.

The outage probability is defined as the probability that end-to-end SNR is dropped below a threshold γ_{th}. Thus, the outage probability for proactive DF relay selection scheme is defined as

$$P_{out}^{Pro} = Pr\{\gamma^{Pro} < \gamma_{th}\} \qquad (9)$$

where $\gamma_{th} = 2^{\frac{2R}{B}} - 1$, where B is the bandwidth and R is the channel rate.

2.2.2 Reactive DF Relay Selection

According to reactive DF relay scheme [15], a set of relays is defined which can successfully decode the incoming signal. Let this set be ξ. Therefore the set can be explained as $\xi = \left\{ \forall k \in (1, 2, \ldots, K); \gamma_{SR_k} \geq \mu_{th} \right\}$ where $\gamma_{SR_k} = \frac{P_s}{N_o}$ is the instantaneous SNR. The decoding process at R_k is successful only when $\gamma_{SR_k} > \mu_{th}$. Here μ_{th} is the threshold SNR for deciding successful decoding at the relays. The instantaneous SNR between the $SS \to SR_k \to SD$ links can be represented as

$$\gamma_{R_k D} = \frac{P_{R_k}}{N_o} \left| h_{R_k D} \right|^2 \tag{10}$$

where P_{R_k} is obtained from Eq. (7).

According to the decoding set ξ, one relay is chosen as [15]

$$\gamma^{Re} = \max_{k \in \xi} \left\{ \gamma_{R_k D} \right\} \tag{11}$$

The outage probability of reactive DF scheme can be represented as

$$P_{Out}^{Re} = Pr\{\gamma^{Re} < \gamma_{th}\} \tag{12}$$

where $\gamma_{th} = 2^{\frac{2R}{B}} - 1$, B is the bandwidth, and R is the channel rate.

3 Simulation Model

In this section, we describe the simulation model that we have developed using MATLAB to evaluate the performance of the considered system model.

- First we have defined the required parameters like I_p, number of antenna (L), number of relays (R), channel rate, μ_{th}.
- Then we have defined a non-identical Rayleigh fading channel for the channel between secondary source to relay, secondary source to primary user, relay to secondary destination, and relay to primary user.
- Additive white Gaussian noise ($AWGN$) samples are added to signal while it is transmitted through channel.
- Transmitted power at source and relay is then calculated using the formula defined in Eqs. (1) and (7), respectively.
- Instantaneous SNR for source to relay and relay to destination is calculated using the transmitted power formulated in the above step.
- Now the end-to-end SNR is individually calculated for both proactive and reactive DF relay selection schemes.

- A condition statement is used wherein the source to relay SNR is compared to the threshold SNR μ_{th} for deciding successful decoding at the relays.
- If the source to relay SNR matches the threshold SNR μ_{th}, then reactive DF selection is done.
- If the source to relay SNR does not match the value of SNR μ_{th}, then transmission is stopped.
- Then the outage probability of reactive DF is calculated taking threshold SNR = 3 dB values, and various performance curves are evaluated.
- The max of min evaluation of end-to-end SNR is done to compute the outage probability for proactive DF scheme.

4 Results and Discussions

A simulation test bed is developed using MATLAB to evaluate the outage performance of the considered system model. The results and their respective analysis are presented below.

Unless stated otherwise, we set the energy conversion efficiency $\eta = 1$, the path loss exponent $n = 2.7$, the distances d_1 and d_2 are set at unit value, the primary user transmit power is set at 10 dB, and the channel rate R is fixed at 1 bits/sec.

The plot shown in Fig. 3 is of outage probability against $\frac{I_p}{N_o}$ for varying μ_{th}. As stated earlier, μ_{th} is the deciding SNR for determining whether the relay has successfully decoded the incoming signal or not. If the source to relay SNR γ_{SR_k} is greater than μ_{th} then the relay has successfully decoded the source signal otherwise not. The reactive relay selection scheme is compared to proactive scheme and shown that for different values of μ_{th} the reactive scheme may generate better or worse results than the proactive scheme. The number of relays K is set at 3, and the number of antennas L is also set at 3. For $\mu_{th} = 3$ dB the reactive outage probability coincides with proactive outage probability. When μ_{th} is less than 3 dB, better results are observed for reactive scheme and when μ_{th} is greater than 3 dB result degradation is seen. Thus, it cannot be concluded whether proactive is better than reactive because the reactive scheme changes its outage performance with changing values of μ_{th} whereas proactive scheme does not depend on μ_{th}.

Figure 4 shows the outage probability against $\frac{I_p}{N_o}$ for proactive and reactive relay selection schemes for increasing number of relays. The number of antennas l is kept constant at 3. The outage performance of both proactive and reactive relaying increases considerably as number of relays are increased from 1 to 5. This is because, as number of relays increases the SNR diversity also increases. The proactive scheme shows better performance because under this system parameter setting, the relays may not be able to decode the incoming signal successfully which leads to bad performance of the reactive scheme.

Figure 5 is outage probability as a function of $\frac{I_p}{N_o}$ for proactive and reactive DF relay selection schemes with fixed number of relays, i.e., $K = 3$ and number of

Fig. 3 Outage performance comparison between proactive and reactive DF relay selection schemes for varying μ_{th}

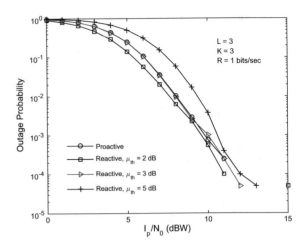

Fig. 4 Comparison of outage performance for proactive and reactive DF relaying schemes for increasing number of relays

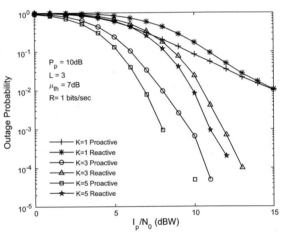

antennas at each relay is varied as $L = 1, 3, 5$. It is seen that for both proactive and reactive schemes, the outage performance increases as number of antennas increases. This is because as the number of antennas are increased from 1 to 5, several copies of same signal reaches the relay. Thus, the probability of selecting the best signal increases considerably and hence the outage performance increases. However no significant performance enhancement is seen prior to 5 antennas at each relay for both proactive and reactive relaying schemes. This is because most of the relays successfully decode the source signal when L is above 5. So, there is no need to increase the number of antennas prior to 5.

Figure 6 shows the outage probability against $\frac{I_p}{N_o}$ for different combinations of number of antennas (L) and number of relays (K) in the considered network. Three combinations are considered—$L = 1, K = 3$, $L = 3, K = 3$, and $L = 3, K = 1$. For the combination $L = 3, K = 3$, the outage performance surpasses the other two

Fig. 5 Comparison of outage performance for proactive and reactive DF relaying schemes for increasing number of antennas

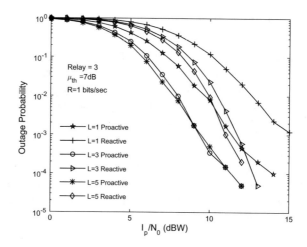

Fig. 6 Comparison of outage performance for proactive and reactive DF relaying schemes for combinations of L and K

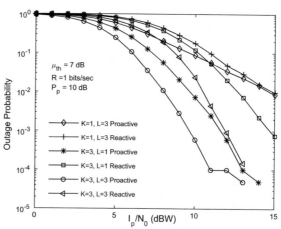

combinations. In first, i.e., $L = 1$, $K = 3$ case, only a single copy of incoming signal is available to each relay which generates high outage probability. In the final case, i.e., $L = 3$, $K = 1$, multiple copies of incoming signal are available to the single relay, which is a major disadvantage because the relay may not be able to decode the signal and also there is no SNR diversity. So both proactive and reactive relaying schemes suffer, thus, degrading the outage performance. In the intermediate situation $L = 3$, $K = 3$, multiple copies of same signal are available to multiple relays which increase the SNR diversity and relay selection probability to a large extent. So, we see a noticeable performance increase in the intermediate case for both proactive and reactive relaying schemes. Thus, we can arrive at a trade-off between number of relays K and number of antennas at each relay L to get desirable outage performance.

5 Conclusion

In this paper, we have compared the outage performance of an energy harvesting underlay CR network using proactive and reactive DF relay selection schemes. *TSR* protocol is followed at the secondary relays for joint work of energy harvesting and signal transmission. It has been observed that the outage probability decreases with increase in number of relays which corresponds to better performance of the network. It is seen that increasing the number of antennas at each relay also increases the performance. A trade-off between number of antennas at each relay and total number of relays is also shown in order to achieve a desired level of performance. It is also seen that the reactive DF scheme may generate better or worse results compared to the proactive DF scheme for varying values of μ_{th} which makes it impossible to decide the better of them.

References

1. J. Mitola, G.Q. Maguire, Cognitive radio: making software radios more personal. IEEE Pers. Commun. **6**(4), 13–18 (1999)
2. M.L. Ku, W. Li, Y. Chen, K.J.R. Liu, Advances in energy harvesting communications: past, present, and future challenges. IEEE Commun. Surv. Tutorials **18**(2), 1384–1412 (2016)
3. X. Lu, P. Wang, D. Niyato, D.I. Kim, Z. Han, Wireless networks with RF energy harvesting: a contemporary survey. IEEE Commun. Surv. Tutorials **17**(2), 757–789 (2015)
4. P. Gandotra, R.K. Jha, S. Jain, Green communication in next generation cellular networks: a survey. IEEE Access **5**, 11727–11758 (2017)
5. M.M. Mowla, I. Ahmad, D. Habibi, Q.V. Phung, A green communication model for 5G systems. IEEE Trans. Green Commun. Netw. (2017)
6. D.B. da Costa, H. Ding, J. Ge, Z. Jiang, Asymptotic analysis of cooperative diversity systems with relay selection in a spectrum-sharing scenario. IEEE Trans. Veh. Technol. **60**(2), 457–472 (2011)
7. J.G. Andrews, J. Lee, H. Wang, D. Hong, Outage probability of cognitive relay networks with interference constraints. IEEE Trans. Wirel. Commun. **10**(2), 390–395 (2011)
8. V.A. Bohara, A. Vashistha, S. Sharma, Outage analysis of a multiple antenna cognitive radio system with cooperative decode-and forward relaying. IEEE Wirel. Commun. Lett. **4**(2), 125–128 (2015)
9. H.Y. Kong, Multi-relay cooperative diversity protocol with improved spectral efficiency. **13**, 240–249 (2011)
10. K. Ho-Van, Exact outage probability analysis of proactive relay selection in cognitive radio networks with MRC receivers. J. Commun. Netw. **18**, 288–298 (2016)
11. P. Yang, J. Song, M. Li, G. Zhang, Outage probability gap between proactive and reactive opportunistic cognitive relay networks. Wirel. Pers. Commun.**82**(4), 2417–2429 (2015). https://doi.org/10.1007/s11277-015-2356-8
12. Y. Bai, Y. Mai, N. Wang, Performance comparison and evaluation of the proactive and reactive routing protocols for manets, in *2017 Wireless Telecommunications Symposium (WTS)*, April 2017, pp. 1–5
13. A.A. Nasir, D.T. Ngo, X. Zhou, R.A. Kennedy, S. Durrani, Joint resource optimization for multicell networks with wireless energy harvesting relays. IEEE Trans. Veh. Technol. **65**(8), 6168–6183 (2016)

14. Y. Gu, S. Aïssa, RF-based energy harvesting in decode-and-forward relaying systems: ergodic and outage capacities. IEEE Trans. Wirel. Commun. **14**(11), 6425–6434 (2015)
15. C.K. De, S. Kundu, Proactive and reactive DF relaying for cognitive network with multiple primary users. Radioengineering **25**(3), 475 (2016)
16. X. Zhou, R. Zhang, C.K. Ho, Wireless information and power transfer: architecture design and rate-energy tradeoff. IEEE Trans. Commun. **61**(11), 4754–4767 (2013)

Butler Matrix Fed Exponentially Tapered H-Plane Horn Antenna Array System Using Substrate Integrated Folded Waveguide Technology

Wriddhi Bhowmik, Vibha Rani Gupta, Shweta Srivastava and Laxman Prasad

Abstract This paper introduces the implementation of 4 × 4 Butler matrix fed antenna array system using substrate integrated folded waveguide (SIFW) technology. The antenna array system generates four directive beams in different angular directions with gain and half power beam width (HPBW) of 6.43, 6.43, 6.33, and 6.43 dB and 12º, 9º, 10º, and 10º. These directive beams will help to suppress the interfering signals from unwanted users, and hence, the quality of high-speed wireless communication will be improved.

Keywords Butler matrix array · Substrate integrated waveguide · Substrate integrated folded waveguide (SIFW) and horn antenna

1 Introduction

Minimization of interference and multipath fading makes to improve the performance of high-speed wireless communication. Multiple beam forming antenna systems (M-BFAS) can be used for this purpose. It generates very directive and narrow beams in different angular directions. These directive beams reduce the possibility of interference as well as fading. The Butler matrix array is the suitable option of beam forming network (BFN) as very few components require for matrix implementation [1]. Also, the matrix introduces low loss to the microwave signal transmission.

W. Bhowmik (✉)
Haldia Institute of Technology, Haldia, West Bengal, India
e-mail: bhowmikwriddhi@gmail.com

V. Rani Gupta
Birla Institute of Technology, Mesa, Ranchi, Jharkhand, India

S. Srivastava
Jaypee Institute of Information Technology, Noida, Uttar Pradesh, India

L. Prasad
Raj Kumar Goel Institute of Technology, Ghaziabad, India

© Springer Nature Singapore Pte Ltd. 2017
J. Bhaumik et al. (eds.), *Communication, Devices, and Computing*, Lecture Notes in Electrical Engineering 470, https://doi.org/10.1007/978-981-10-8585-7_3

The Butler matrix array comprises of basic microwave devices, such as 3 and 0 dB coupler, 45° phase shifter. Previously, the implementation of Butler matrix array has been made by microstrip technique to make the system compact as well as inexpensive. Most of the times researchers used rectangular patch microstrip antenna as radiating element in M-BFAS [2–6], to generate directive beams in different angular directions. Besides these advantages, the microstrip feeding network suffers from undesired feed radiation as well as encountered with high ohmic and dielectric losses at higher frequencies. These lead to the degradation of radiation performances.

To overcome these difficulties, waveguide technology has been used for the development of Butler matrix array [7]. In [7], the improved performance has been achieved at the price of costly and voluminous configuration, due to the non-planar construction of waveguide.

The substrate integrated waveguide (SIW) technology transforms non-planar configuration of waveguide into the planar form. The planar implementation of waveguide Butler matrix array has been obtained by SIW technology [8–11].

To make the system more compact, substrate integrated folded waveguide technology (SIFW) has been introduced in this paper. The SIFW technique reduces the width of SIW by a factor of half as well as the individual building blocks. The size reduction of 3 and 0 dB coupler and 45° phase shifter will make the overall system more compact. Polytetrafluoroethylene (PTFE) with the relative permittivity (ε_r) of 2.5, height (h) of 0.5 mm, and dissipation factor (tan δ) of 0.00015 has been used for design implementation. To design microwave devices, it is required to use high dielectric constant. On the other side, material with lower relative permittivity is used for antenna design. The proposed structure contains both the microwave devices and antenna. Moreover, the structure has been realized by using planar waveguide technology (SIFW technique). Hence, the performance of Butler matrix network is quite satisfactory, though the dielectric constant is low. In this proposed work, exponentially tapered SIFW H-plane horn antenna has been used as radiating element as it has narrow beam width.

The realization of beam forming network using SIFW technique to obtain compactness as well as the achievement of narrow directive beams by the physical integration of exponentially tapered SIFW H-plane horn antenna with the output ports of BFN is unique. The generated directive beams will very efficiently reject the interfering signal and improve the quality of communication.

1.1 Butler Matrix Array

Generally, a Butler matrix array [12, 13] is a $N \times N$ matrix ($N = 2^n$). The output ports of Butler matrix array feed N element array produce N directive beams in different angular directions. Distinct phase differences are obtained between the output ports of the matrix, while input ports get excited by a signal. The direction of the radiated beams depends upon these phase differences. The beams direction θ can be calculated as follows [14]

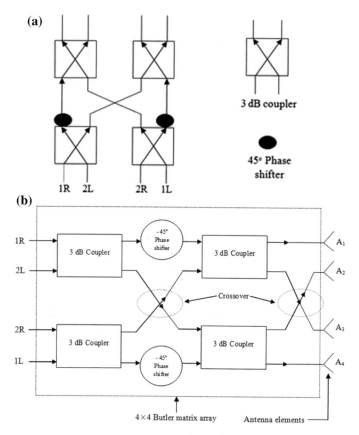

Fig. 1 **a** Schematic of 4 × 4 Butler matrix array and **b** M-BFAS

$$\theta = \cos^{-1} \left[\frac{\pm (2n - 1) \lambda_0}{8d} \right] \tag{1}$$

In Eq. (1), $n = 1, 2\ldots$ etc., λ_0 is wavelength, and d is the center to center distance of antenna elements.

Figure 1a and Figure 1b, respectively, represents the schematic of 4 × 4 Butler Matrix array and multiple beam forming antenna system (M-BFAS).

The power will flow through the Butler matrix network, while any input port of the matrix gets excited and finally one-fourth of the input power fed to the antenna elements. The cumulative effect of 4 element array (shown in Fig. 1b) will produce a directive beam. The excitation of input port named 1R (1st Right) of M-BFAS (shown in Fig. 1b) yields the directive beam 1R, which should be in the angular range of 0° to 30°, as shown in Fig. 2. Similarly, the 1L, 2R, and 2L beams should be in the angular range of −30° to 0°, 30° to 60°, and −60° to −30°, respectively, while the ports 1L, 2R, and 2L of M-BFAS get excited.

Fig. 2 Proposed radiation pattern M-BFAS

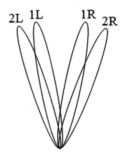

Fig. 3 A simple structure of SIW

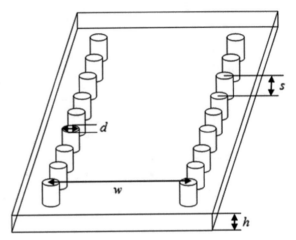

1.2 Substrate Integrated Waveguide and Substrate Integrated Folded Waveguide

The substrate integrated waveguide (SIW) technology makes it possible to fabricate the rectangular waveguides in a planar form. The classical rectangular waveguides can easily be integrated with the planar structures using this technology. It is possible to preserve the advantages of rectangular waveguides such as high Q factor and high power handling capability by the SIW technique. SIW technology reduces the losses associated with the microstrip lines. The implementation of SIW structure in the planar form is possible by using periodic metallic via holes, airholes, or holes filled with a different dielectric. The SIW made with airholes or holes filled with a different dielectric produces more leakage in comparison to the SIW with metallic via holes. The metallic pins shield the electromagnetic waves. Figure 3 shows a simple SIW structure formed by arrays of metallic vias.

Fig. 4 Design of SIFW

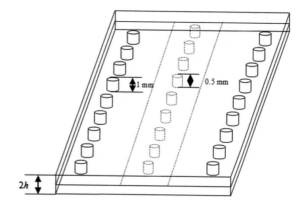

The width (w) of SIW can be calculated as follows [15, 16]

$$w_{eff} = w - 1.08\frac{d^2}{s} + 0.1\frac{d^2}{w} \tag{2}$$

In Eq. (2), d is the diameter of metallic vias, s is the periodicity of the vias, and w_{eff} is the effective width of equivalent rectangular waveguide. To minimize the leakage loss of SIW structure, the ratio of s to d (s/d) should be less than 2.5 as well as d to w (d/w) should be less than 1/5. For TE$_{10}$ mode, the w_{eff} can be calculated as follows [17]

$$w_{eff} = \frac{c}{2\,(f_c)_{10}\,\sqrt{\mu_r \varepsilon_r}} \tag{3}$$

In Eq. (3), $c = 3 \times 10^8$ m/s, μ_r and ε_r are the relative permeability and permittivity of dielectric material, and $(f_c)_{10}$ is the cutoff frequency of the waveguide for TE$_{10}$ mode.

Size reduction factor of rectangular waveguide by SIW technique is $1/\sqrt{\varepsilon_r}$. Again, it is possible to reduce the size of SIW by half with the help of SIFW technique by keeping the cutoff frequency unchanged [18, 19]. The structure of SIFW is shown in Fig. 4.

2 Design of SIFW H-Plane Exponentially Tapered Horn Antenna

The proposed horn antenna will be integrated with the output ports of SIFW Butler matrix array to generate directive beams. Parametric study has been carried out on the dimensions of SIW horn antenna presented in [20], as per the operating frequency to get satisfactory result. The size reduction of the aperture of SIW H-plane horn

Fig. 5 Proposed SIFW exponentially tapered H-plane horn antenna: **a** top layer and **b** middle layer

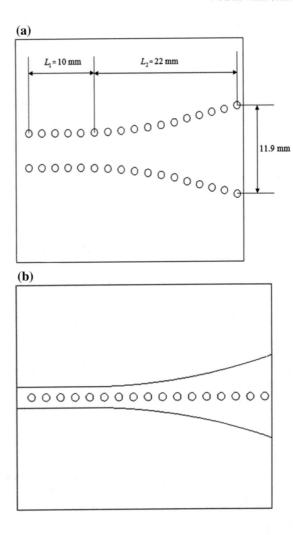

antenna [20, 21] has been achieved by SIFW technique. For a given aperture of horn, improvement of radiation performance might be achieved by increasing the horn length as well as decreasing the flaring angle. To meet those above requirements, exponential tapering has been used. The aperture length and flaring angle are approximately 31 mm and 21.7°, respectively. The proposed horn antenna is a two-layer structure, and it is presented in Fig. 5.

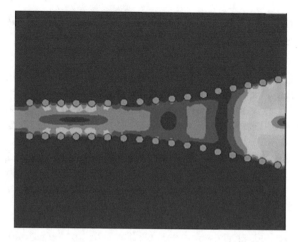

Fig. 6 Simulated current distribution at 15 GHz

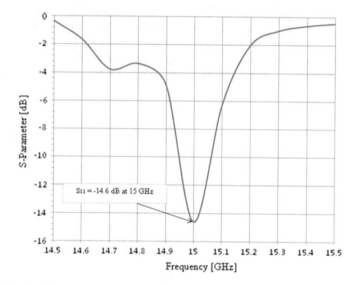

Fig. 7 Simulated S-parameter characteristic

The current distribution of the proposed horn antenna at 15 GHz is presented in Fig. 6. The S_{11} of −14.6 dB has been achieved by the proposed design as shown in Fig. 7. The H-plane gain pattern of SIFW horn antenna has been reported in Fig. 8. A gain of 2.61 dB is obtained at 15 GHz. The small size of the aperture is responsible for the back radiation.

3   Design of SIFW M-BFAS Using Exponentially Tapered H-Plane Horn Antenna

The realization of proposed BFN comprises of successful integration of SIFW 3 and 0 dB double slot coupler and 45° phase shifter. Successful integration of all microwave devices leads to satisfactory performances of the system at the frequency of interest. The proposed structure of SIFW Butler matrix array is presented in Fig. 9.

The exponentially tapered H-plane horn antenna, presented in Sect. 2, is integrated with the output ports (P_5, P_6, P_7, and P_8) of Butler matrix array to generate the directive beams in different angular directions. Figure 10a, b presents the structure of proposed M-BFAS.

Good impedance matching has been obtained in simulation while input ports of M-BFAS get excited, as shown in Fig. 11. The ratios presented in Fig. 11 depict that, most of the input power is transmitted to the SIFW M-BFAS and less amount of power has been reflected back. The normalized H-plane radiation patterns of the proposed M-BFAS for the different input port excitations (1R, 2L, 2R, and 1L) at 15 GHz are presented in Fig. 12a, Fig. 12b, Fig. 12c, and Fig. 12d, respectively. The radiated beams are very directive with reduced side lobe levels of about −20 dB as shown in Fig. 12. The radiated major beams (1R, 2L, 2R, and 1L) directions are 12°, −30°, 30°, and −12°, respectively.

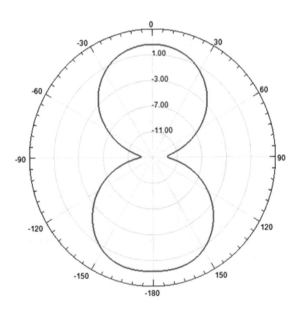

Fig. 8 Simulated H-plane radiation pattern

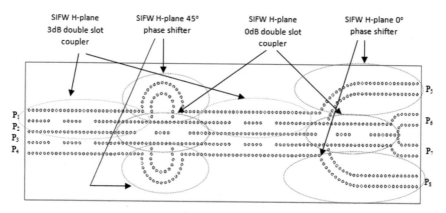

Fig. 9 Layout of proposed SIFW Butler matrix array

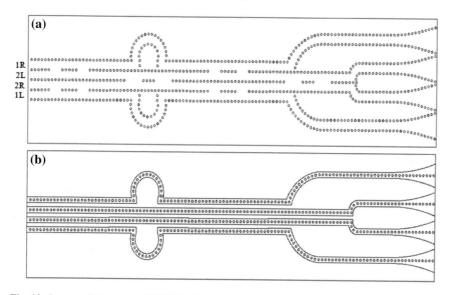

Fig. 10 Layout of the proposed SIFW M-BFAS: **a** top layer and **b** middle layer

Directivity and gain of radiated beams have been reported in Table 1. The reported data projects that symmetric gain and directivity have been obtained for all radiated beams at 15 GHz. The half power beam widths of 12°, 9°, 10°, and 10° have been obtained for the beams 1R, 2L, 2R, and 1L, respectively. Hence, it shows that very narrow beams have been achieved for different port excitations.

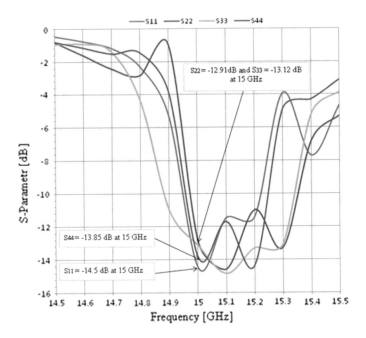

Fig. 11 Simulated S_{11}, S_{22}, S_{33}, and S_{44} for 1R, 2L, 2R, and 1L ports excitation, respectively

Table 1 Directivity and gain of the proposed SIFW M-BFAS at 15 GHz

Beams	Directivity (dB)	Gain (dB)
1R	6.81	6.43
2L	6.93	6.43
2R	6.81	6.33
1L	6.81	6.43

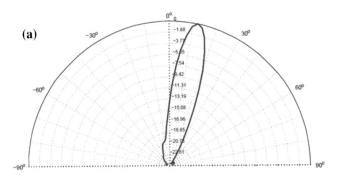

Fig. 12 Simulated radiation performances of the proposed SIFW M-BFAS at 15 GHz for different ports excitation: **a** 1R; **b** 2L; **c** 2R; and **d** 1L

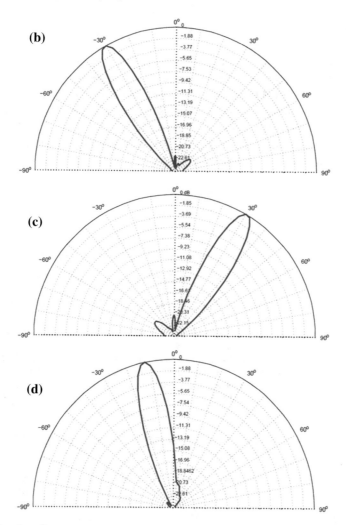

Fig. 12 (continued)

4 Conclusion

Individual building blocks of Butler matrix array as well as the BFN and M-FAS have been realized by using SIFW technology. The normalized H-plane radiation performances of the SIFW M-BFAS have been simulated. Very directive and narrow beams with reduced side lobe levels and HPBW have been achieved by the proposed system. Hence, the directive beams can be used to receive the main signal strength of the mobile users and can efficiently suppress the interfering signals, and this leads to improve the quality of high-speed wireless communication.

References

1. W. White, Pattern limitations in multiple-beam antennas. IRE Trans. Antennas Propag. **10**, 430–436 (1962)
2. J.-S. Neron, G.-Y. Delisle, Microstrip EHF Butler matrix design and realization. ETRI J. **27**, 788–797 (2005)
3. O.U. Khan, Design of X-band 4 × 4 Butler matrix for microstrip patch antenna array. TENCON 2006, Hong Kong (2006), pp. 1–4
4. S.Z. Ibrahim, M.K.A. Rahim, Switched beam antenna using omnidirectional antenna array, in *Asia Pacific Conference on Applied Electromagnetics*, Malaysia (2007), pp. 1–4
5. A.M. El-Tager, M.A. Eleiwa, Design and implementation of smart antenna using Butler matrix for ISM-band, in *Progress in Electromagnetic Research Symposium*, China (2009), pp. 571–575
6. C.-H. Tseng, C.-J. Chen, T.-H. Chu, A low-cost 60-GHz switched-beam patch antenna array with Butler matrix network. IEEE Antennas Wirel. Propag. Lett. **07**, 432–435 (2008)
7. J. Remez, R. Carmon, Compact designs of waveguide Butler matrices. IEEE Antennas Wirel. Propagat. Lett. **05**, 27–31 (2006)
8. T. Djerafi, N.J.G. Fonseca, K. Wu, Design and implementation of a planar 4 × 4 Butler matrix in SIW technology for wide band high power applications. Prog. Electromagn. Res. B. **35**, 29–51 (2011)
9. C.-J. Chen, T.-H. Chu, Design of a 60-GHz substrate integrated waveguide Butler matrix-a systematic approach. IEEE Trans. Microw. Theory Tech. **58**, 1724–1733 (2010)
10. A.A.M. Ali, N.J.G. Fonseca, F. Coccetti, H. Aubert, Design and implementation of two-layer compact wideband Butler matrices in SIW technology for Ku-band applications. IEEE Trans. Antennas Propag. **59**, 503–512 (2011)
11. Y.J. Cheng, C.A. Zhang, Y. Fan, Miniaturized multilayer folded substrate integrated waveguide Butler matrix. Prog. Electromagn. Res. C. **21**, 45–58 (2011)
12. J. Butler, R. Lowe, Beam-forming matrix simplifies design of electrically scanned antennas. Electron. Design. **9**, 170–173 (1961)
13. S.R. Ahmad, F.C. Seman, 4-port Butler matrix for switched multibeam antenna array, in *Asia Pacific Conference on Applied Electromagnetics*, Malaysia (2005)
14. G.E. Dominguez, J.-M. Fernandez-Gonzalez, P. Padilla, M. Sierra-Castafier, Mutual coupling reduction using EBG in steering antennas. IEEE Antennas Wirel. Propag. Lett. **11**, 1265–1268 (2006)
15. Y. Cassivi, L. Perregrini, P. Arcioni, M. Bressan, K. Wu, G. Conciauro, Dispersion characteristics of substrate integrated rectangular waveguide. IEEE Microw. Wirel. Compon. Lett. **12**, 333–335 (2001)
16. F. Xu, K. Wu, Guided-wave and leakage characteristics of substrate integrated waveguide. IEEE Trans. Microw. Theory Tech. **53**, 66–73 (2005)

17. H. Nam, T.-S. Yun, K.-B. Kim, K.-C. Yoon, J.-C. Lee, Ku—band transition between microstrip and substrate integrated waveguide (SIW), in *Asia-Pacific Microwave Conference*, China (2005)
18. N. Grigoropoulos, B. Sanz-Izquierdo, P.R. Young, Substarte integrated folded waveguides (SIFW) and filters. IEEE Microw. Wirel. Compon. Lett. **15**, 829–831 (2005)
19. N. Grigoropoulos, P.R. Young, Compact folded waveguides, in *34th European Microwave Conference*, Amsterdam (2004), pp. 973–976
20. H. Wang, D.G. Fang, B. Zhang, W.Q. Che, Dielectric loaded substrate integrated waveguide H-plane horn antennas. IEEE Trans. Antennas Propag. **58**, 640–647 (2010)
21. W. Che, B. Fu, P. Yao, Y.L. Chow, E.K.N. Yung, A compact substrate integrated waveguide H-plane horn antenna with dielectric arc lens. Int. J. RF Microw. Comput.-Aided Eng. 473–479 (2007)

Computing Characteristic Impedance of MIM Nano Surface Plasmon Structure from Propagation Vector Characteristics for Skin Depth Measurement

Pratibha Verma and Arpan Deyasi

Abstract Characteristic impedance of MIM surface plasmon structure is analytically computed from the knowledge of propagation vector variation for different nanometric structural dimensions. Length, width, and distance between metal plates are kept in nanometer range to evaluate the propagation vector in high frequency domain, and effect of both Faraday inductance and kinetic inductance are incorporated. Notch in impedance profile is observed due to the modulation of surface plasmon propagation vector w.r.t dielectric propagation vector, and position of notch can suitably be tailored by proper choice of dimensions and material. Result provides accurate indirect measurement of the skin depth of the structure.

Keywords Characteristic impedance · Propagation vector · MIM structure
Surface plasmon · Structural parameters · Dielectric constant

1 Introduction

Surface plasmon phenomenon, which is nothing but the oscillation of conductive electrons, is occurred at the interface of two different media in which one has negative permittivity and other has positive permittivity [1], when longitudinal excitations arise from the conductive electron gas is collected [2]. This oscillation is very sensitive to the properties of the boundary, i.e., the nature of the adjoining surfaces. Devices under resonant condition is already used to screen drug candidates [3], generate PDMS (poly-di-methyl-siloxane) microchips [4], designing new generation of optic and nanoelectronic devices [5], simultaneous detection of biological analytes

P. Verma
Department of Electronics and Communication Engineering, NIT Agartala,
Jirania, India
e-mail: vermapratibha1007@gmail.com

A. Deyasi (✉)
Department of Electronics and Communication Engineering,
RCC Institute of Information Technology, Kolkata, India
e-mail: deyasi_arpan@yahoo.co.in

© Springer Nature Singapore Pte Ltd. 2017
J. Bhaumik et al. (eds.), *Communication, Devices, and Computing*, Lecture Notes
in Electrical Engineering 470, https://doi.org/10.1007/978-981-10-8585-7_4

[6] etc. In such cases, estimation of propagation vector becomes really important as it governs the dispersion relation [7, 8]. Spatial extent and coupling requirement for insight matching of the SP structure is obtained from the knowledge of propagation vector [9].

In this present work, propagation vector of the nanometric surface plasmon structure is computed for different dimensions of metal–insulator–metal (MIM) structure, and therefore, characteristic impedance is obtained. Here Ag–air–Ag is considered (Ag has negative permittivity at operating frequency). In Sect. 2, detailed analytical calculation is presented considering effect of kinetic and Faraday inductances and dependence of propagation vector on material permittivity and skin depth is shown. Result reveals that skin depth of the structure can be determined for given operating frequency once propagation vector is estimated.

2   Mathematical Modeling

We know that Faraday inductance

$$L_f = \frac{\mu_o d}{W}(Length) \tag{1}$$

and kinetic inductance

$$L_k = \frac{1}{\omega^2 \varepsilon_o(1 - \varepsilon_r'(\omega))} \frac{(Length)}{(Area)} \tag{2}$$

where 'Length' and 'Area' are dependent on the structure concerned. Propagation vector for MIM structure is a combination of them as

$$k = \omega\sqrt{\frac{L_k + L_f}{C}} \tag{3}$$

Substituting the values, we get

$$k = \omega\left[\frac{\frac{1}{\omega^2 \varepsilon_o(1-\varepsilon_r'(\omega))} \frac{(Length)}{(Area)} + \frac{\mu_o d}{W}(Length)}{\varepsilon_0 \frac{W}{d}}\right]^{\frac{1}{2}} \tag{4}$$

Rearranging, one can obtain

$$k = \omega\left[\frac{length \times d\left(1 + \delta\mu_0 d\omega^2\varepsilon_0\left(1 - \varepsilon_r'(\omega)\right)\right)}{W^2\delta\omega^2\varepsilon_0^2\left(1 - \varepsilon_r'(\omega)\right)}\right]^{\frac{1}{2}} \tag{5}$$

Permittivity of the metal is given by

$$\varepsilon_r = (n + ik_{ex})^2 \tag{6}$$

where 'n' is the refractive index of the metal and 'k_{ex}' is the extinction coefficient of the metal.

Separating real part and substituting Eq. (5), we get

$$\varepsilon'_r = \left(\frac{c\omega}{2\pi\lambda}\right)^2 - \frac{\alpha c}{2\omega} \tag{7}$$

where α is the absorption coefficient.

As real part of propagation vector is dependent on dielectric constants of metal and insulator, as well as frequency of operation, given by

$$k = \sqrt{\left[\frac{\omega}{c}\left(\frac{\varepsilon'_r \varepsilon_d}{\varepsilon'_r + \varepsilon_d}\right)^{1/2}\right]^2 + \left[\frac{\omega}{c}\left(\frac{\varepsilon'_r \varepsilon_d}{\varepsilon'_r + \varepsilon_d}\right)^{3/2}\frac{\varepsilon''_r}{2\left(\varepsilon'_r\right)^2}\right]^2} \tag{8}$$

where ε_d is the permittivity of the insulator of the MIM structure.

Equating (5) with (8), we obtain

$$\left[\frac{\omega^2}{c^2}\left(\frac{\varepsilon'_r \varepsilon_d}{\varepsilon'_r + \varepsilon_d}\right)\right] + \left[\frac{\omega^2}{c^2}\left(\frac{\varepsilon'_r \varepsilon_d}{\varepsilon'_1 + \varepsilon_d}\right)^3\frac{\left(\varepsilon''_r\right)^2}{4\left(\varepsilon'_r\right)^4}\right] =$$
$$\omega^2\left[\frac{length \times d\left(1 + \delta\mu_0 d\omega^2\varepsilon_0(1 - \varepsilon'_r(\omega))\right)}{W^2\delta\omega^2\varepsilon_0^2\left(1 - \varepsilon'_r(\omega)\right)}\right] \tag{9}$$

From Eq. (9), permittivity of the metal can be determined through dimension of the MIM structure and skin depth for given operating frequency.

Characteristic impedance for nanometer range of MIM structure for lossless condition is

$$Z = \sqrt{\frac{k^2}{\omega^2} - \frac{L_f}{C}} \tag{10}$$

Then 'Z' can be modified as

$$Z = \sqrt{\frac{k^2}{\omega^2} - \frac{\mu_0 d^2}{W^2\delta\varepsilon_0}length} \tag{11}$$

For lossy medium,

$$Z = \sqrt{\frac{R + j\omega(L_k + L_f)}{G + j\omega C}} \qquad (12)$$

Substituting the values of passive components, we get

$$Z = \sqrt{\frac{\frac{\varepsilon_r''(\omega)}{\omega\varepsilon_0|1-\varepsilon_r'(\omega)|^2}\frac{\text{Length}}{W\delta} + j\omega\left(\frac{1}{\omega^2\varepsilon_0(1-\varepsilon_r'(\omega))}\frac{(\text{Length})}{(W\delta)} + \frac{\mu_o d}{W}(\text{Length})\right)}{\sigma\frac{W\delta}{\text{Length}} + j\omega\varepsilon_0\frac{W\delta}{d}}} \qquad (13)$$

Substituting the value of skin depth $\delta = \sqrt{\frac{2}{\omega\sigma\mu_0}}$, final expression of characteristic impedance is given by

$$Z = \sqrt{\frac{\frac{\varepsilon_r''(\omega)}{\omega\varepsilon_0|1-\varepsilon_r'(\omega)|^2}\frac{\text{Length}}{W\sqrt{\frac{2}{\omega\sigma\mu_0}}} + j\omega\left(\frac{1}{\omega^2\varepsilon_0(1-\varepsilon_r'(\omega))}\frac{(\text{Length})}{\left(W\sqrt{\frac{2}{\omega\sigma\mu_0}}\right)} + \frac{\mu_o d}{W}(\text{Length})\right)}{\sigma\frac{W\sqrt{\frac{2}{\omega\sigma\mu_0}}}{\text{Length}} + j\omega\varepsilon_0\frac{W\delta}{d}}} \qquad (14)$$

3   Results and Discussions

Using Eq. (8) and Eq. (14), propagation vector and characteristic impedance of the plasmon structure are calculated, respectively. In Fig. 1, propagation vector is calculated as a function of different structural parameters. For electromagnetic radiation, the wavenumber is proportional to the energy of the radiation. Hence, the more the magnitude of propagation vector, the higher the radiation of energy of electromagnetic wave passing through the MIM structure. In Fig. 1a, propagation vector is plotted with distance between metal plates for different lengths, where simulation is made for Ag. At 633 nm, permittivity of the sliver is -18.295; hence plasmon is created between the boundaries of air and metal. For thinner metal in nm range, electron energy is highly lost or radiated for effect of surface plasmon, and henceforth, momentum of bulk plasmon is raised which causes increase in propagation vector.

Capacitive value is step down with higher distance between two metal plates (d) of MIM structure, and it drives the reason to store of energy in lesser amount which varies inversely proportional to 'd'. Thus, radiation of energy is enhanced and proportionally k is also increased. Oscillation of electron gas along 'k' is in more area for higher value of 'L' and large surface is occupied for the radiation of energy with larger value of 'L'. Thereby, 'k' is also increased. Moreover, kinetic inductance (L_k) and Faraday inductance (L_f) induced in the MIM structure are directly proportional to the length of metal plates and hence propagation vector is directly proportional to the summation of L_k and L_f. Thus, allover 'k' is increased with increasing of length.

Fig. 1 **a** Propagation vector profile with distance between metal plates for different lengths, **b** propagation vector with distance between metal plates for different metals

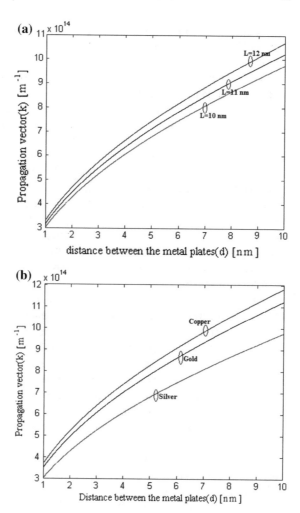

Variation of distance between the metal plates (d) w.r.t propagation vector (k) for different metals is indicated in Fig. 1b, where magnitude of 'k' is least for silver and is highest for copper. Real values of relative permittivity of copper, gold, and silver are −11.56, −11.753, and −18.295, respectively, at wavelength of 633 nm with their corresponding conductivity as 5.85×10^7 S m^{-1}, 4.87×10^7 S m^{-1}, 6.17×10^7 S m^{-1}. 'k' is inversely proportional to relative permittivity of the metal plate but directly proportional to the square root of conductivity of the metal. Hence, skin depth of penetration of the electromagnetic field is highest in gold and least in silver. Therefore, it can be concluded that 'k' increases w.r.t the increase in distance between two metal plates where the distance is measured as highest in case of copper and least in silver.

In Fig. 2a, variation of characteristic impedance (Z) w.r.t the propagation vector (k) is depicted where impedance is initially decreased with increase of propagation vector. But after cutoff range, Z is monotonically increased. Characteristic impedance of a medium is the ratio of amplitude of voltage and current for a single wave propagating along the medium which is determined by its geometry and material characteristic. As 'k' is high, radiation of energy along the wave propagation is also high that causes less confinement of the energy and decreases amplitude of the voltage. This phenomenon is responsible for fall of characteristic impedance up to certain range of 'k'. But as the 'k' is increased, surface plasmon (SP) increases as magnitude of component of 'k' along SP mode is greater than magnitude of propagation vector of dielectric medium. As long as propagation vector to the SP is greater than propagation vector to the dielectric, energy of SP mode will be confined and will not leak because of evanescent coupling between the SP in the top surface and the plane waves of the dielectric in the bottom surface so that the energy of SP is leaking as the plane waves toward the bottom dielectric. Hence, amplitude of the voltage is enhanced comparatively and causes to increase of 'Z' with magnification of 'k' after certain cutoff.

Figure 2a and Figure 2b exhibit the impedance variation with different 'L' and different 'd', respectively. Characteristic impedance is inversely proportional to conduction current and directly proportional to kinetic inductance. Thus, 'Z' is enhanced with increase of 'L' for certain range of 'k'. But it is also noticed that after certain range of increment of 'k', impedance is step up due to more confinement of energy of SP mode. As length is increased, there is more oscillation of electron gas that occurs from surface plasmon phenomenon and so, there is increment in conduction current. Thus, Z falls down with step up of conduction current. Therefore, Z is decreased with increase of length of metal plates of MIM structure after certain cutoff range of 'k'.

Characteristic impedance falls down as thickness of insulating material of MIM structure 'd' is decreased because capacitance between the two metal plates of MIM structure is increased. As 'd' is decreased, the presence of the plasmon on one interface is seen on other interface, and the two SP modes between the two metal plates and dielectric are more coupled which causes to more confinement of energy with small loss and hence, increases of amplitude of voltage of the wave comparatively. Thus, 'Z' is increased with reduction of 'd' after certain cutoff range of 'k'.

In Fig. 2c, variation of characteristic impedance with propagation vector is plotted for different metal layers. It is seen from the plot that 'Z' is highest for Ag in certain range of propagation vector because in this range penetration depth δ is dominant. 'Z' is inversely proportional to 'δ' and hence directly proportional to conductivity of the metal. But, after this cutoff range of 'k', when 'Z' increases with increase of 'k' then SP mode is dominant that causes to more oscillation of electron and conduction current gets higher due to higher conductivity of the metal and also more oscillation of electron. Thus, 'Z' is decreased with increase of conduction current and increase of conductivity of the metal. So, in second range of enhancement of propagation vector, the characteristic impedance is raised for lower value of conductivity and this is the reason here that 'Z' is lowest in silver and highest in gold.

Fig. 2 **a** Characteristic impedance profile with propagation vector for different lengths for Ag–air–Ag system, **b** characteristic impedance profile with propagation vector for different 'd' for Ag–air–Ag system, **c** characteristic impedance profile with propagation vector for different metals

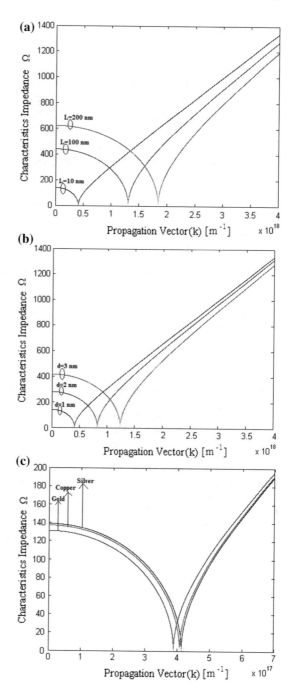

4 Conclusion

Characteristic impedance of the nano surface plasmon structure is analytically calculated as a function of propagation vector after evaluating the later as a function of different structural parameters and metal permittivity. Computation is performed considering the lossy nature of the structure, and effect of both Faraday and kinetic inductances are incorporated. Result exhibits the modulation in impedance profile, and the magnitude of corresponding propagation vector responsible for zero impedance. Simulated findings will help to compute skin depth for the structure under different structural configurations.

References

1. J. Homola, S.S. Yee, G. Gauglitz, Surface plasmon resonance sensors: review. Sens. Actuators B **54**, 3–15 (1999)
2. A.K. Sharma, R. Jha, B.D. Gupta, Fiber-optic sensors based on surface plasmon resonance: a comprehensive review. IEEE Sens. J. **7**, 1118–1129 (2007)
3. C. Boozer, G. Kim, S. Cong, H. Guan, T. Londergan, Looking towards label-free biomolecular interaction analysis in a high-throughput format: a review of new surface plasmon resonance technologies. Curr. Opin. Biotechnol. **17**, 400–405 (2006)
4. J.D. Taylor, M.J. Linman, T. Wilkop, Q. Cheng, Regenerable tethered bilayer lipid membrane arrays for multiplexed label-free analysis of lipid-protein interactions on poly(dimethylsiloxane) microchips using SPR imaging. Anal. Chem. **81**, 1146–1153 (2009)
5. C. Prasad, S.K. Shrivastav, L.K. Mishra, A theoretical evaluation of quality factor of surface plasmon-polaritons (Q_{SPP}) as a function of wavelength (nm) and comparative studies of metals for better plasmonic materials. J. Chem. Biol. Phys. Sci. Sect. C **6**(2), 546–555 (2016)
6. B. Hong, A. Sun, L. Pang, A.G. Venkatesh, D. Hall, Y. Fainman, Integration of Faradaic electrochemical impedance spectroscopy into a scalable surface plasmon biosensor for in tandem detection. Opt. Express **23**(23), 30237–30249 (2015)
7. A.V. Zayats, I.I. Smolyaninov, A.A. Maradudin, Nano-optics of surface plasmon polaritons. Phys. Rep. **408**(3), 131–314 (2005)
8. J.M. Pitarke et al., Theory of surface plasmons and surface-plasmon polaritons. Rep. Prog. Phys. **70**(1), 1 (2006)
9. D. Courjon, K. Sarayeddine, M. Spajer, Scanning tunneling optical microscopy. Opt. Commun. **71**(1–2), 23–28 (1989)

Extended Directional IPVO for Reversible Data Hiding Scheme

Sudipta Meikap and Biswapati Jana

Abstract Pixel Value Ordering (PVO) is a data hiding technique through which pixels within an image block are ranked and modify minimum or maximum pixel value for data embedding. In this paper, we proposed an Extended Directional Improved PVO (EDIPVO) for Reversible Data Hiding (RDH) scheme. The original image pixel blocks are interpolated by inserting extra pixels between each row and each column. During data embedding, we consider a parameter (α), perform addition with maximum pixel value, and subtract α from minimum pixel value to maintain the order within the pixel block. The parameter α is dependent on the size of the original image block. To increase data hiding capacity, we have considered pixel ranking in three different directions: (1) horizontal, (2) vertical, and (3) diagonal within each block. The secret data bits are embedded starting from the largest, second largest, and so on for maximum value modification and smallest, second smallest, and so on for minimum value modification. The data embedding capacity has been improved than previous data hiding schemes while keeping visual quality unaltered. Experimental results are compared with existing state-of-the-art methods and achieve good results.

Keywords Reversible data hiding · Pixel-value-ordering · Prediction-error expansion · Embedding capacity · Steganography · PSNR · EDIPVO

1 Introduction

Innocent data communication through steganography is increasing day by day due to the demanding issue of online exchange of valuable information. Reversible Data Hiding (RDH) is a process of reconstruction or recovery original image after data

S. Meikap (✉)
Department of Computer Science, Hijli College, Paschim Midnapore
721306, West Bengal, India
e-mail: sudiptameikap@gmail.com

B. Jana
Department of Computer Science, Vidyasagar University, Midnapore 721102, India
e-mail: biswapatijana@gmail.com

© Springer Nature Singapore Pte Ltd. 2017
J. Bhaumik et al. (eds.), *Communication, Devices, and Computing*, Lecture Notes in Electrical Engineering 470, https://doi.org/10.1007/978-981-10-8585-7_5

extraction which is desirable in many application areas such as medical image processing, military application, remote sensing and judicial application for copyright protection, authentication, and tampered detection. The RDH schemes are evaluated by two important parameters, such as data hiding capacity and visual distortion. The main challenge of an efficient RDH scheme is to improve visual quality by reducing pixel modification for a given amount of secret data.

The early data hiding scheme was initiated by modifying LSB bits in cover image introduced by Turner [1] in 1989. There was asymmetry which can be overcome by Sharp [2] in 2001. To exchange LSB-based data hiding scheme, Milikainen [3] proposed good quality stegos. Recently, RDH schemes attracted much attention to the researchers. Tian [4] designed a data hiding scheme using difference expansion (DE) method to hide secret bits within a pair of pixel. Alattar [5] modified Tian's scheme and used the difference between four pixels. Lee et al. [6] utilized histogram of the difference of pixel values to hide secret data within cover image for improving the visual quality of stego image. An RDH scheme using histogram shifting has been suggested by Ni et al. [7]. Later, Thodi et al. [8] presented RDH scheme that combines histogram shifting and difference expansion. After that, Lin et al. [9] and Tsai et al. [10] proposed improved RDH scheme through multilevel histogram shifting.

In 2013, Li et al. [11] proposed data hiding through PVO scheme where pixels are ordered and modify largest value to hide maximum secret data. Peng et al. [12] improve both embedding capacity and visual quality where more smooth blocks are used. Challenges to improve embedding capacity are possible by considering repetition the embedding process in different direction within a block. So, we take some initiative to improve Peng et al. [12] scheme by repeated embedding on marked pixels. Here, we consider data hiding in horizontal, vertical, and diagonal directions one after another. In each direction, the secret data bits are embedded by computing the largest, second largest, and so on for maximum modification and smallest, second smallest, and so on for minimum modification. The embedding capacity has been increased than Peng et al. scheme, whereas the quality is unchanged.

2   Proposed Method

In this section, we describe data embedding and extraction procedure of proposed Extended Directional Improved PVO (EDIPVO). The data embedding can take place in Peng et al. [12], by modification of minimum and maximum pixel values of image block. We try to improve data embedding capacity by maintaining good visual quality through our proposed Extended Directional Improved PVO (EDIPVO) scheme. In this scheme, we try to embed hidden message in first, second, third, fourth, and so on with a pair of minimum and maximum pixel values of row / column / diagonal for interpolated image block. The total procedure is categorized in the following steps.

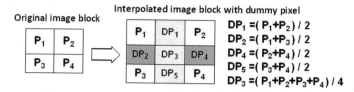

Fig. 1 Block diagram of interpolated image block. Red and yellow color represent interpolated row and column, respectively

2.1 Data Embedding Procedure

Step 1: At first, we enlarge original image block using Image Interpolation technique shown in Fig. 1. The interpolated row and column depend on the size of original block. For example, if the size of original block is $(m \times m)$, then the interpolated image block will be $(m + (m - 1)) \times (m + (m - 1))$, where $(m - 1)$ is number of interpolated row and column in interpolated image block.

Step 2: Now, data embedding procedure will be done on horizontal direction of interpolated image block. In this technique, it will be possible to modify more than two pixel values. So, there will be a chance to change ascending order sorting position. To make same order, we subtract a parameter α from first, second, third, and so on minimum pixel value and add α to first, second, third, and so on maximum pixel value. The value of α will depend upon the size of original pixel block. If m is the block size of original pixel block, then $(2m - 1)$ is the interpolated image block. So, the maximum value of α will be $(m - 2)$ and minimum value of α will be 0.

Lemma 1 *If the interpolated pixel block size is $(2m - 1)$ and the ascending order sorted pixel value in a row / column / diagonal is $(p_1, p_2, p_3, \ldots, p_{2m-3}, p_{2m-2}, p_{2m-1})$, then the modified pixel value will be $(p_1 - \alpha_{(m-2)}, p_2 - \alpha_{((m-2)-1)}, p_3 - \alpha_{((m-2)-2)}, \ldots, p_{m-1} - \alpha_{((m-2)-(m-2))}, p_m, p_{m+1} + \alpha_{((m-2)-(m-2))}, \ldots, p_{2m-3} + \alpha_{((m-2)-2)}, p_{2m-2} + \alpha_{((m-2)-1)}, p_{2m-1} + \alpha_{(m-2)}), where the value of \alpha_{((m-2)-(m-2))} = ((m - 2) - (m - 2)), maximum value of \alpha = (m - 2), and minimum value of \alpha = 0.*

Example 1 If the original block size $(m) = 4$, then interpolated image block will be (7×7). The maximum value of α is $(m - 2) = 2$, which is subtracted from minimum and added to the maximum pixel value. The next value of α is $(m - 2) - 1 = 1$, which is subtracted from second minimum pixel value and added to the second maximum pixel value. Then, the next value of α is $(m - 2) - 1 - 1 = 0$, which is subtracted from third minimum pixel value and added to the third maximum pixel value. Now, the decrement process of α is stopped, because it reaches to $(m - 2) - 1 - 1 = 0$. The pixels of interpolated image block are shown in Fig. 2.

Now, secret data will be embedded into pairs of minimum and maximum pixel values depending upon the original image block size according to the procedure given below.

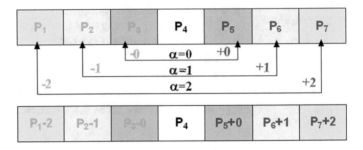

Fig. 2 Block diagram of example of Lemma 1

2.2 Minimum Modification-Based Data Embedding

The details of Minimum Modification-Based Data Embedding are described as follows: Now, we sort pixel values in ascending order to obtain $(x_{\sigma(1)}, \ldots, x_{\sigma(n)})$, where $\sigma : \{1, \ldots, n\} \to \{1, \ldots, n\}$ and $x_{\sigma(1)} \leq \ldots \leq x_{\sigma(n)}$, $\sigma(i) < \sigma(j)$ if $x_{\sigma(i)} = x_{\sigma(j)}$ and $i < j$ is one to one mapping of a given block X containing n pixels $(x_1, \ldots, x_n)$. Now, we calculate

$$d_{\min_t} = x_a - x_b \text{ where } \begin{cases} a = \min(\sigma((2) + t), \sigma((1) + t)), \\ b = \max(\sigma((2) + t), \sigma((1) + t)), \\ t = (0, 1, 2, \ldots, fix(n/2) - 1). \end{cases} \quad (1)$$

- If $\sigma((2) + t) > \sigma((1) + t0$, then $a = \sigma((1) + t)$ and $b = \sigma((2) + t)$. For this case, $d_{\min_t} \leq 0$.
- If $\sigma((2) + t) < \sigma((1) + t)$, then $a = \sigma((2) + t)$ and $b = \sigma((1) + t)$. For this case, $x_{\sigma((2)+t)} > x_{\sigma((1)+t)}$ and $d_{\min_t} > 0$.

Now, the difference $d_{\min_t}$ is modified to $d'_{\min_t}$ which is obtained by the following equation

$$d'_{\min_t} = \begin{cases} d_{\min_t} - 1, & \text{if } d_{\min_t} < 0 \\ d_{\min_t} - B, & \text{if } d_{\min_t} = 0 \\ d_{\min_t} + B, & \text{if } d_{\min_t} = 1 \\ d_{\min_t} + 1, & \text{if } d_{\min_t} > 1 \end{cases} \quad (2)$$

where bit $B \in \{0, 1\}$ which is embedded to the pixel value. In each time, the value of α is modified and the minimum value $x_{\sigma((1)+t)}$ will be modified as x' which is obtained by

$$x' = x_{\sigma((2)+t)} - \alpha - |d'_{\min_t}| = \begin{cases} (x_{\sigma((1)+t)} - \alpha) - 1, & \text{if } d_{\min_t} < 0 \\ (x_{\sigma((1)+t)} - \alpha) - B, & \text{if } d_{\min_t} = 0 \\ (x_{\sigma((1)+t)} - \alpha) - B, & \text{if } d_{\min_t} = 1 \\ (x_{\sigma((1)+t)} - \alpha) - 1, & \text{if } d_{\min_t} > 1 \end{cases} \quad (3)$$

Let the marked value of X be $(y_1, y_2, \ldots, y_n)$, where $y_{\sigma((1)+t)} = x'$ and $y_i = x_i$ for every $i \neq \sigma((1) + t)$. Let the marked value of X be $(R_1, R_2, \ldots, R_n)$ in column-wise, where $R_{\sigma((1)+t)} = r'$ and $R_i = r_i$ for every $i \neq \sigma((1) + t)$.

2.3 Maximum Modification-Based Data Embedding

The details of Maximum Modification-Based Data Embedding are described below. We calculate

$$d_{\max_t} = x_c - x_d \text{ where } \begin{cases} c = \min(\sigma((n-1) - t), \sigma((n) - t)), \\ d = \max(\sigma((n-1) - t), \sigma((n) - t)), \\ t = (0, 1, 2, \ldots, fix(n/2) - 1). \end{cases} \quad (4)$$

- If $\sigma((n-1) - t) < \sigma((n) - t)$, then $c = \sigma((n-1) - t)$ and $d = \sigma((n) - t)$. For this case, $d_{\max_t} \leq 0$.
- If $\sigma((n-1) - t) > \sigma((n) - t)$, then $c = \sigma((n) - t)$ and $d = \sigma((n-1) - t)$. For this case, $x_{\sigma((n-1)-t)} < x_{\sigma((n)-t)}$ and $d_{\max_t} > 0$.

Now, the difference $d_{\max_t}$ is modified to $d'_{\max_t}$ which is obtained by

$$d'_{\max_t} = \begin{cases} d_{\max_t} - 1, & \text{if } d_{\max_t} < 0 \\ d_{\max_t} - B, & \text{if } d_{\max_t} = 0 \\ d_{\max_t} + B, & \text{if } d_{\max_t} = 1 \\ d_{\max_t} + 1, & \text{if } d_{\max_t} > 1 \end{cases} \quad (5)$$

where bit $B \in \{0, 1\}$ which is embedded to the maximum pixel value. In each time, value of α will be modified and the maximum pixel value $x_{\sigma((n)-t)}$ will be modified as x' which is obtained as

$$x' = x_{\sigma((n-1)-t)} + \alpha + |d'_{\max_t}| = \begin{cases} (x_{\sigma((n)-t)} + \alpha) + 1, & \text{if } d_{\max_t} < 0 \\ (x_{\sigma((n)-t)} + \alpha) + B, & \text{if } d_{\max_t} = 0 \\ (x_{\sigma((n)-t)} + \alpha) + B, & \text{if } d_{\max_t} = 1 \\ (x_{\sigma((n)-t)} + \alpha) + 1, & \text{if } d_{\max_t} > 1 \end{cases} \quad (6)$$

Let the marked value of X be $(y_1, y_2, \ldots, y_n)$, where $y_{\sigma((n)-t)} = x'$ and $y_i = x_i$ for every $i \neq \sigma((n) - t)$. Let the marked value of X be $(R_1, R_2, \ldots, R_n)$ in column-wise, where $R_{\sigma((n)-t)} = r'$ and $R_i = r_i$ for every $i \neq \sigma((n) - t)$. Thus, data embedding procedure on horizontal direction is completed and after getting the marked image block from horizontal direction, the above said embedding procedure is occurred on vertical direction. Finally, after getting the marked image block from vertical direction, the said embedding procedure is occurred on diagonal direction.

2.4 Data Extraction Procedure

At receiver side, we perform data extraction procedure from stego image. We first perform extraction procedure in diagonal direction on marked image block. We add and subtract the value of α to minimum pixel value and from maximum pixel value of first, second, third and so on respectively, depend upon the following Lemma 2.

Lemma 2 *If the marked pixel block size is $(2m - 1)$ and the ascending order sorted pixel value in a row / column / diagonal is $(p_1', p_2', p_3', \ldots, p_{2m-3}', p_{2m-2}', p_{2m-1}')$, then the modified pixel value will be $(p_1' + \alpha_{(m-2)}, p_2' + \alpha_{((m-2)-1)}, p_3' + \alpha_{((m-2)-2)}, \ldots, p_{m-1}' + \alpha_{((m-2)-(m-2))}, p_m', p_{m+1}' - \alpha_{((m-2)-(m-2))}, \ldots, p_{2m-3}' - \alpha_{((m-2)-2)}, p_{2m-2}' - \alpha_{((m-2)-1)}, p_{2m-1}' - \alpha_{(m-2)})$, where the value of $\alpha_{((m-2)-(m-2))} = ((m - 2) - (m - 2))$, maximum value of $\alpha = (m - 2)$, and minimum value of $\alpha = 0$.*

Example 2 If marked image block is (7×7), then the original block size$(m) = 4$. The maximum value of α is $(m - 2) = 2$, which is added to minimum and subtracted from the maximum pixel value. The next value of α is $(m - 2) - 1 = 1$, which is added to second minimum pixel value and subtracted from the second maximum pixel value. Then, the next value of α is $(m - 2) - 1 - 1 = 0$, which is added to third minimum pixel value and subtracted from the third maximum pixel value. Now, the decrement process of α is stopped, because it reaches to $(m - 2) - 1 - 1 = 0$. So, the pixels of marked image block are shown in Fig. 3.

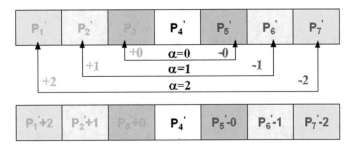

Fig. 3 Block diagram of example of Lemma 2

2.5 Minimum Modification-Based Data Extraction

At the decoder side, it can perform the data extraction procedure and finally image restoration process from the marked image block has been done where the marked value is $(y_1, y_2, \ldots, y_n)$. Here, the mapping σ remains unchanged. We calculate $d'_{\min_t} = y_a - y_b$, where (a, b, t) is defined in Eq. (1).

- If $d'_{\min_t} \leq 0$, then $y_a \leq y_b$. Now, $a = \sigma((1) + t)$, $b = \sigma((2) + t)$, and $\sigma((1) + t) < \sigma((2) + t)$:

  - if $d'_{\min_t} \in \{0, -1\}$, then hidden data is present and data bit $B = -d'_{\min_t}$ and the restored minimum pixel value is $x_{\sigma((1)+t)} = (y_a + \alpha) + B$;
  - if $d'_{\min_t} < -1$, then there is not present hidden data and the restored minimum pixel value is $x_{\sigma((1)+t)} = (y_a + \alpha) + 1$.

- If $d'_{\min_t} > 0$, then $y_a > y_b$. Now, $a = \sigma((2) + t)$, $b = \sigma((1) + t)$, and $\sigma((1) + t) > \sigma((2) + t)$:

  - if $d'_{\min_t} \in \{1, 2\}$, then hidden data is present and the data bit $B = d'_{\min_t} - 1$ and the recovered minimum pixel value is $x_{\sigma((1)+t)} = (y_b + \alpha) + B$;
  - if $d'_{\min} > 2$, then there is not present hidden data and the original minimum pixel value is $x_{\sigma((1)+t)} = (y_b + \alpha) + 1$.

2.6 Maximum Modification-Based Data Extraction

At the decoder side, perform data extraction procedure and finally original image restoration process from the marked image block has been done where the marked value is $(y_1, y_2, \ldots, y_n)$. Here, the mapping σ remains unchanged. Now, we calculate $d'_{\max_t} = y_c - y_d$ where (c, d, t) is defined previously in Eq. (4).

- If $d'_{\max_t} \leq 0$, then $y_c \leq y_d$. Now, $c = \sigma((n-1) - t)$, $d = \sigma((n) - t)$, and $\sigma((n-1) - t) < \sigma((n) - t)$:

  - if $d'_{\max_t} \in \{0, -1\}$, then hidden data is present and the data bit $B = -d'_{\max_t}$ and the recovered maximum pixel value is $x_{\sigma(n)-t} = (y_d - \alpha) - B$;
  - if $d'_{\max_t} < -1$, then there is not present hidden data and the restored maximum pixel value is $x_{\sigma((n)-t)} = (y_d - \alpha) - 1$.

- If $d'_{\max_t} > 0$, then $y_c > y_d$. Now, $c = \sigma((n) - t)$, $d = \sigma((n-1) - t)$, and $\sigma((n-1) - t) > \sigma((n) - t)$:

  - if $d'_{\max_t} \in \{1, 2\}$, then hidden data is present and the data bit $B = d'_{\max_t} - 1$ and the original maximum pixel value is $x_{\sigma((n)-t)} = (y_c - \alpha) - B$;
  - if $d'_{\max_t} > 2$, then there is not present hidden data and the recovered maximum pixel value is $x_{\sigma((n)-t)} = (y_c - \alpha) - 1$.

Thus, the hidden data is recovered and pixel values are restored from all diagonal of marked image. Then, the above said procedure executes for vertical direction and next horizontal direction on marked image block. The hidden massage is recovered from each image block. All the interpolated rows and columns are eliminated from each interpolated marked image block and restored original image block.

3 Experimental Result

In this section, the performance of our proposed method is compared with existing PVO-based data hiding developed by Li et al. [11], Peng et al. [12], Ou et al. [13], and Qu et al. [14]. We have taken ten grayscale images with size (512 × 512) including Lena, Airplane F 16, Sailboat on lake, Peppers, Baboon, Tiffany, Barbara, Fishing boat, Elaine, and House that are used in our experiment as test images which are shown in Fig. 4. All images are collected from the USC-SIPI database. The data embedding and data extraction algorithms are tested using MATLAB 7.6.0.324 (R2008a).

The embedding performance is depended on the size of original block which is shown in Fig. 5. It is observed that the performance is better in larger block size compared with the smaller block size but capacity is lower. For example, 1,18,986 bits are embedded within Lena image (512 × 512) for (2 × 2) block size but only 43,646 bits are embedded for (8 × 8) block size.

Fig. 4 The standard test images are used for our experiments

Fig. 5 Embedding performance of EDIPVO method with different block sizes (for Lena image)

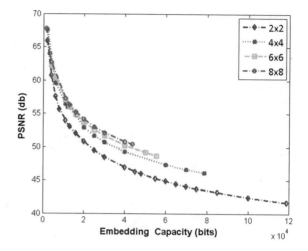

Table 1 Comparison in terms of PSNR (dB) for the methods of Li et al. [11], Peng et al. [12], Ou et al. [13], Qu et al. [14] and the proposed method with the payload of 10,000 bits

Image	Li et al. [11]	Peng et al. [12]	Ou et al. [13]	Qu et al. [14]	Proposed
Lena	59.8	60.4	60.6	60.3	58.8
Airplane F 16	61.6	62.9	63.3	63.7	61.5
Sailboat on lake	58.2	58.8	59.4	59.8	57.1
Peppers	58.5	58.9	59.2	58.8	57.8
Baboon	53.5	53.5	54.5	54.2	52.8
Tiffany	60.1	60.7	60.3	60.6	59.7
Barbara	59.9	60.5	60.6	59.8	59.1
Fishing boat	57.8	58.2	58.1	58.4	56.2
Elaine	56.8	57.3	57.4	58.7	56.1
House	61.8	64.4	63.7	64.6	63.7
Average	58.8	59.5	59.7	59.8	58.2

Our proposed method increases the embedding capacity (EC) compared with existing PVO-based data hiding scheme. Our EC is 86,986, 80,986, 81,986, 72,986 bits greater than Li et al., Peng et al., Ou et al. and Qu et al. EC respectively for Lena image Fig. 6 and Tables 1 and 2.

Table 2 Comparison of embedding capacity (EC) in number of bits

Image	Li et al. [11]	Peng et al. [12]	Ou et al. [13]	Qu et al. [14]	Proposed
Lena	32,000	38,000	37,000	46,000	1,18,986
Airplane F 16	38,000	52,000	47,000	69,000	1,70,334
Sailboat on lake	23,000	26,000	26,000	29,000	80,974
Peppers	28,000	30,000	31,000	33,000	95,312
Baboon	13,000	13,000	13,000	15,000	42,632
Tiffany	33,000	40,000	40,000	52,000	1,39,618
Barbara	27,000	29,000	31,000	33,000	1,30,056
Fishing boat	24,000	26,000	26,000	30,000	80,504
Elaine	21,000	24,000	23,000	29,000	81,302
House	30,000	46,000	37,000	64,000	1,55,622
Average	26,900	32,400	31,100	40,000	1,09,534

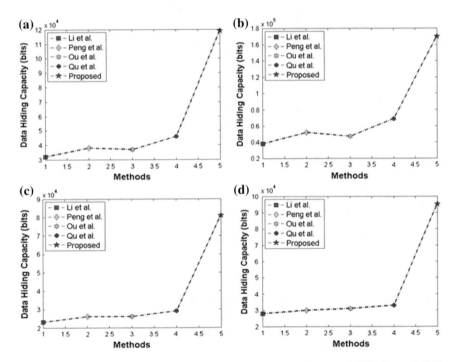

Fig. 6 Comparisons of EC among the methods of Li et al. [11], Peng et al. [12], Ou et al. [13], Qu et al. [14] and our proposed method

4 Conclusion

In this paper, we proposed an extension of IPVO scheme for RDH. The main goal is to communicate secret data by modifying minimum and maximum value pixel in the block. Peng et al. [12] suggested embedding data by considering largest and second largest for maximum modification and smallest and second smallest for minimum modification using only one direction without any repetitions. To increase embedding capacity through generalized PVO-based data hiding scheme, we have introduced a parameter α which is dependent on the size of the original pixel block. During data embedding, parameter (α) is added with maximum pixel value and subtract α from minimum pixel value to maintain the pixel order within the block. As a result, we can increase the size of the original image block without disturbing their order and it is possible to embed secret data using parameter α. We have proposed EDIPVO for varying block size (i.e., for generalization) and taking three different directions: (1) horizontal, (2) vertical, and (3) diagonal within each block. The secret data bits are embedded starting from the largest, second largest, and so on for maximum modification and smallest, second smallest, and so on for minimum modification. Extensive experiments verified that the proposed method outperformed than the state-of-the-art works.

References

1. L.F. Turner, Digital data security system. Pat. IPN wo **89**, 08915 (1989)
2. T. Sharp, An implementation of key-based digital signal steganography, in *International Workshop on Information Hiding* (Springer, Berlin, Heidelberg, April 2001), pp. 13–26
3. J. Mielikainen, LSB matching revisited. IEEE Signal Process. Lett. **13**(5), 285–287 (2006)
4. J. Tian, Reversible data embedding using a difference expansion. IEEE Trans. Circuits Syst. Video Technol. **13**(8), 890–896 (2003)
5. A.M. Alattar, Reversible watermark using the difference expansion of a generalized integer transform. IEEE Trans. Image Process. **13**(8), 1147–1156 (2004)
6. S.K. Lee, Y.H. Suh, Y.S. Ho, *Lossless Data Hiding Based on Histogram Modification of Difference Images, Advances in Multimedia Information Processing-PCM 2004* (Springer, Berlin Heidelberg, 2004), pp. 340–347
7. Z. Ni, Y.Q. Shi, N. Ansari, W. Su, Reversible data hiding. IEEE Trans. Circuits Syst. Video Technol. **16**(3), 354–362 (2006)
8. D.M. Thodi, J.J. Rodríguez, Expansion embedding techniques for reversible watermarking. IEEE Trans. Image Process. **16**(3), 721–730 (2007)
9. C.C. Lin, W.L. Tai, C.C. Chang, Multilevel reversible data hiding based on histogram modification of difference images. Pattern Recognit. **41**(12), 3582–3591 (2008)
10. W.L. Tai, C.M. Yeh, C.C. Chang, Reversible data hiding based on histogram modification of pixel differences. IEEE Trans. Circuits Syst. Video Technol. **19**(6), 906–910 (2009)
11. X. Li, J. Li, B. Li, B. Yang, High-fidelity reversible data hiding scheme based on pixel-value-ordering and prediction-error expansion. Signal Process. **93**(1), 198–205 (2013)

12. F. Peng, X. Li, B. Yang, Improved PVO-based reversible data hiding. Digit. Signal Process. **25**, 255–265 (2014)
13. B. Ou, X. Li, Y. Zhao, R. Ni, Reversible data hiding using invariant pixel-value-ordering and prediction-error expansion. Signal Process. Image Commun. **29**(7), 760–772 (2014)
14. X. Qu, H.J. Kim, Pixel-based pixel value ordering predictor for high-fidelity reversible data hiding. Signal Process. **111**, 249–260 (2015)

Hamming Code-Based Watermarking Scheme for Image Authentication and Tampered Detection

Pabitra Pal, Partha Chowdhuri, Biswapati Jana and Jaydeb Bhaumik

Abstract In this paper, a (7, 4) Hamming code-based color image watermarking scheme has been proposed for image authentication and tampered detection. Here, watermark image is generated after embedding watermark symbol within a color cover image. The cover image is partitioned into (7 × 7) pixel blocks and separated into three R, G, B color blocks. To embed watermark, (7, 4) Hamming code is used in each block. For each watermark embedding operation, 21 bits of watermark information are embedded within each color image block. The proposed scheme extracts secret data successfully from watermarked image. The authentication and tampered detection have been tested on watermarked image. Finally, the proposed scheme has been compared with the related existing state of the art methods in terms of visual quality.

Keywords Watermarking · Hamming code · Matrix coding · Payload
PSNR · Standard deviation and correlation coefficient

1 Introduction

In last two decades, the art of secret writing and communication of information have been marked a significant role due to the development of communication technology. Watermarking is a technique to embed secret information within the object which is

P. Pal · P. Chowdhuri · B. Jana (✉)
Department of Computer Science, Vidyasagar University, Midnapore 721102,
West Bengal, India
e-mail: biswapatijana@gmail.com

P. Pal
e-mail: pabipaltra@gmail.com

P. Chowdhuri
e-mail: prc.email@gmail.com

J. Bhaumik
Department of Electronics and Communication Engineering, HIT, Midnapore 721657,
West Bengal, India
e-mail: bhaumik.jaydeb@gmail.com

© Springer Nature Singapore Pte Ltd. 2017
J. Bhaumik et al. (eds.), *Communication, Devices, and Computing*, Lecture Notes
in Electrical Engineering 470, https://doi.org/10.1007/978-981-10-8585-7_6

one of the promising solutions for tampered detection, copyright protection, owner identification and authentication of digital content. However, irreversible watermarking creates some problem which damages the original information that presents in the cover work. But it is highly desirable in many areas like military communication, healthcare, online publishing, remote sensing, and law enforcement. In such applications, reversible watermarking is required.

In 1950, R. W. Hamming [1] first introduced the error detecting and correcting codes. In 1998, matrix coding-based data hiding technique [2] was introduced by Crandall where they embed k bits information into $(2^k - 1)$ cover work. In their scheme, they achieve $k/(2^k + 1)$ bpp payload. After that in 2001, Westfeld modified the matrix coding technique and proposed F5 algorithm [3]. In 2006, Fridrich and Soukal [4] presented a modified data hiding technique through matrix embedding method and structure code to achieve high embedding capacity, which was 0.9 bpp. Next Zhang et al. [5] modified the Crandall's scheme by combining Hamming codes and covering code to maximizing the embedding efficiency and they increase the payload to $(k + 1)/(2^k)$ bpp. After that, they modified their technique in [6] to acquire a good balance between efficiency and capacity, in addition to wet paper codes. Then Chang et al. [7] developed a data hiding process through (7, 4) Hamming Code to achieve 0.99 bpp payload with average PSNR value 50 dB. In 2011, Kim et al. [8] suggested a new embedding technique in a Halftone Image Using (15, 11) Hamming Code, which gives 0.25 bpp payload. After that, Kim and Yang [9] proposed another security-based data hiding scheme named "Hamming+3" technique which gives 0.86 bpp payload with 48 dB PSNR. To increase the payload with good visual quality, Mohamed and Mohamed in [10] presented a new Steganographic technique based on LSB Substitution Method and they achieve 2 bpp payload with 46.59 dB PSNR. Jana et al. [11] proposed a Hamming code-based data embedding technique which one is partially reversible. It provides average embedding payload 0.14 bits per pixel (bpp) with good visual quality which is higher than 59.53 dB. So far, Hamming code-based watermarking scheme has been developed for grayscale image and few schemes are available which are reversible. Also, the embedding capacity and visual quality are limited, when using Hamming code-based watermarking schemes.

Thus, there is a lot of scopes to increase embedding capacity and enhance visual quality through Hamming code-based watermarking techniques which are still an important area of current research. Also, image authentication and tampered detection is a demanding issue for online hidden data communication. So, our objective is to design a secure watermarking scheme which provides authentication in case tampered image.

The paper organization is mentioned below: Proposed scheme is described in Sect. 2. Then, Sect. 3 presented experiments and comparison with the existing scheme. In Sect. 4 security analysis is provided and some security attacks have been performed in Sect. 5. Finally, conclusions are drawn in Sect. 6.

2 Proposed Watermarking Scheme

2.1 Watermark Embedding

In this paper, a color image watermarking technique through $(7, 4)$ Hamming code has been proposed.

The algorithmic explanation to embed the watermark into the cover image is given below:

Step-1:   Consider color image (C) as cover media.
Step-2:   Partition into (7×7) pixel blocks.
Step-3:   Separate three different R, G, B color blocks.
Step-4:   Take $R_{(7 \times 7)}$ of the first block.
Step-5:   Perform odd parity adjustment to the redundant bits.
Step-6:   Take a shared secret key lambda (λ).
Step-7:   Embed the 7 bit of watermark (w_i) into the corresponding row of $R_{(7 \times 7)}$ block by error creation.
Step-8:   Repeat Step-5 to Step-7 using $G_{(7 \times 7)}$ and $B_{(7 \times 7)}$ blocks also.
Step-9:   Select remaining (7×7) blocks and repeat Step-4 to Step-8.
Step-10:   Update the image matrix according to the modified color block.
Step-11:   Generate watermarked image (C').

A numerical illustration is shown in Fig. 1.

2.2 Watermark Extraction

The details watermark extraction algorithm of our proposed method is as follows:

Step-1:   Consider watermarked image (C').
Step-2:   Partition into (7×7) pixel blocks.
Step-3:   Separate three different R, G, B color blocks.
Step-4:   Take $R_{(7 \times 7)}$ of the first block.
Step-5:   Convert the block into LSB matrix.
Step-6:   Complement the bit in the shared secret key (λ) position.
Step-7:   Get the data by Hamming error correction.
Step-8:   7 bits watermark (w_i) are extracted from each row of the $R_{(7 \times 7)}$ block.
Step-9:   Complement the error bit and update the secret location by error position.
Step-10:   Repeat Step-5 to Step-9 using $G_{(7 \times 7)}$ and $B_{(7 \times 7)}$ blocks.
Step-11:   Select next block and repeat Step-4 to Step-9.
Step-12:   Update the image matrix according to the modified color block.
Step-13:   Generate original cover image (C).

The numerical illustration is shown in Fig. 2.

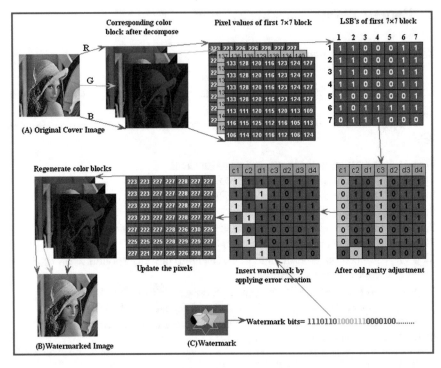

Fig. 1 Numerical illustration of watermark embedding process

Fig. 2 Standard original color images of size (512 × 512) pixels

3 Experiment and Comparison

Here, our proposed watermark scheme is verified using the color image of size (512 × 512) pixels collected from [12] shown in Figs. 2 and 3 shows the generated watermarked images with maximum embedded watermark bits (111,909 bits) within the cover image. The watermark embedding and extraction algorithms are implemented using Java-8. The parameters for measuring the quality and capacity are $PSNR$ and bpp [11] which is given below.

$$MSE = \frac{\sum_{i=1}^{p} \sum_{j=1}^{q} [X(i,j) - Y(i,j)]^2}{(p \times q)}, \tag{1}$$

$$PSNR = 20 \, log_{10} \frac{(255)}{\sqrt{MSE}}, \tag{2}$$

After embedding 111, 909 bits, we achieve nearer 53dB PSNR. To assess the embedding capacity, we used apply following equation.

$$B = \frac{Number \; of \; bits \; embedded}{Total \; number \; of \; pixel \; in \; cover \; image} \tag{3}$$

Here, the payload B of our watermarking scheme is 0.428 bpp.

To measure the complexity, we consider that the size of the cover image is $(p \times q)$ and the watermarking process embeds 21 watermark bits within a (7×7) pixel block.

Fig. 3 Generated watermarked images after embedding watermark within color cover images

Table 1 Comparison with Hamming+1, Nearest Code(NC), Hamming+3, Cao et al. Chang et al. and the proposed scheme in terms of PSNR

Cover image	Hamming+1 [5]	Nearest Code [13]	Hamming+3 [9]	Cao et al. [14]	Chang et al. [7]	Proposed scheme
Lena	52.43	47.02	48.22	51.14	50.84	53.40
Baboon	53.71	…	48.18	…	50.95	53.39
Tiffany	47.46	47.03	48.20	51.15	50.11	53.14
Barbara	48.60	47.01	48.22	51.15	50.36	53.38
Aeroplane	51.61	…	48.20	…	50.77	53.40
Boat	49.37	…	48.20	…	50.39	53.41
Zelda	54.04	…	48.21	…	50.94	53.38
Pepper	47.26	47.02	48.20	51.14	50.12	53.37
Avg. PSNR	50.56	47.02	48.20	51.15	50.56	53.36

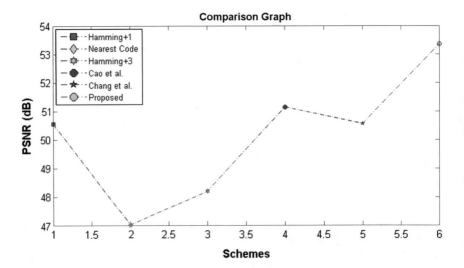

Fig. 4 Comparison graph in terms of PSNR

So, the time complexity for embedding procedure is $\mathcal{O}(pq)$. On the other hand, during extraction, we need to scan pixels from marked image depending on the key. So, the time complexity for extraction procedure is also $\mathcal{O}(pq)$.

Table 1 represents the comparison of the Hamming+1 [5], Nearest Code(NC) [13], Hamming+3 [9], Cao et al. scheme [14], Chang et al. scheme [7] with our proposed scheme in terms of PSNR. It shows that the proposed technique gives better results in terms of imperceptibility which is 53.36 dB PSNR shown in Fig. 4.

Table 2 Standard deviation and correlation coefficient of cover image and watermarked image

Cover image	SD of cover image (C)	SD of watermarked image (C')	CC between C & C'
Lena	128.2929	128.2933	0.9999
Baboon	129.0957	129.1019	0.9999
Tiffany	78.5602	78.4032	0.9999
Barbara	144.1874	144.1947	0.9999
Aeroplane	124.4947	124.5102	0.9999
Boat	191.0677	191.0695	0.9999
Car	149.5743	149.5738	0.9999
Parrot	177.9450	177.8902	0.9999
Zelda	115.0754	115.0892	0.9999
Pepper	136.5673	136.4978	0.9999

4 Steganalysis

To measure the perceptibility, we calculated standard deviation (σ) and correlation coefficient (ρ) for watermarked images which are given in Table 2. We conclude that σ of original image is 128.2929 and the σ of watermark image is 128.2933, and their difference is 0.0004 for Lena image. The ρ between the image C and C' is 0.9999 for Lena image which implies the change in the original image will predict a change in the same direction of watermark image. So, it is hard to locate the embedding position within watermark image. The change in original image is not much defected in watermarked image, which preserves quality and innocuousness. So, the proposed scheme is a perfectly secured watermarking scheme.

5 Attacks

Figure 5 depicted some experimental results for tampered detection from watermarked image using our proposed algorithm. Three types of experiments are presented to show the tampered detection and recovery of the original image. The first one is used 20% tampered image through salt and pepper noise. After performing data extraction, we observed that it has been successfully detected that the watermarked image has tampered and particular watermarked was not present within that tampered watermarked image. As a result, the sender can verify the authentication of marked image. The second one is used 20% tampered with opaque. It is shown that the hidden watermark has not been recovered using data extraction and tampered cover image is recovered. The difference between SD and CC is 1.84 and 0.96, respectively. Finally, we have tested the images tampered by 20% cropping. Here, it is also shown that the recover hidden watermark and cover image, but the quality

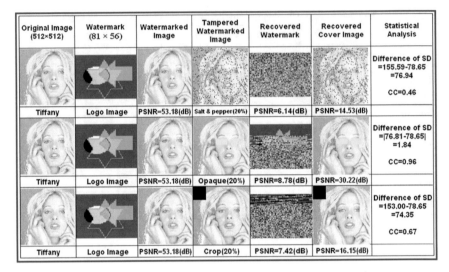

Original Image (512×512)	Watermark (81 × 56)	Watermarked Image	Tampered Watermarked Image	Recovered Watermark	Recovered Cover Image	Statistical Analysis
Tiffany	Logo Image	PSNR=53.18(dB)	Salt & pepper(20%)	PSNR=6.14(dB)	PSNR=14.53(dB)	Difference of SD =155.59-78.65 =76.94 CC=0.46
Tiffany	Logo Image	PSNR=53.18(dB)	Opaque(20%)	PSNR=8.78(dB)	PSNR=30.22(dB)	Difference of SD =|76.81-78.65| =1.84 CC=0.96
Tiffany	Logo Image	PSNR=53.18(dB)	Crop(20%)	PSNR=7.42(dB)	PSNR=16.15(dB)	Difference of SD =153.00-78.65 =74.35 CC=0.67

Fig. 5 Results of watermark extraction from tampered watermarked image of proposed scheme

is not up to the mark. From this experiment, it is mentioned that the adversary cannot extract the watermark from the watermarked image through various tampering approach. But the owner can easily identify and verify the originality of the cover media. If the hidden watermark is not present within any marked object, then we can say that the object has tampered, otherwise it can successfully be verified. So, this scheme may be used for image authentication ownership identification, copyright protection, and tampered detection scheme.

6 Conclusion

In this paper, a watermarking on color image employing Hamming code has been presented. Higher PSNR value establishes that the visual quality of watermarked image is better than the other existing schemes. A shared secret key has been used at the time of embedding and extraction to enhance the security. The effectiveness of proposed scheme against statistical attacks is analyzed. Standard deviation and correlation coefficient values establish the robustness of proposed scheme. The proposed scheme successfully detects the tampered of a marked image through verification of hidden watermarks. The scheme has been used for authentication, tampered detection, ownership identification, and copyright protection. Any educational institute can verify their documents like, mark-sheets, admit cards, certificates, etc, after embedding watermark in each documents.

References

1. R.W. Hamming, Error detecting and error correcting codes. Bell Labs Tech. J. **29**(2), 147–160 (1950)
2. R. Crandall, *Some Notes on Steganography. Steganography Mailing List* (1998)
3. A. Westfeld, F5—a steganographic algorithm, in *International Workshop on Information Hiding, April 2001* (Springer, Berlin, Heidelberg, 2001), pp. 289–302
4. J. Fridrich, D. Soukal, Matrix embedding for large payloads. IEEE Trans. Inf. Forensics Secur. **1**(3), 390–395 (2006)
5. W. Zhang, S. Wang, X. Zhang, Improving embedding efficiency of covering codes for applications in steganography. IEEE Commun. Lett. **11**(8), 680–682 (2007)
6. W. Zhang, X. Zhang, S. Wang, Maximizing steganographic embedding efficiency by combining Hamming codes and wet paper codes, in *International Workshop on Information Hiding, May 2008* (Springer, Berlin, Heidelberg, 2008), pp. 60–71
7. C.C. Chang, T.D. Kieu, Y.C. Chou, A high payload steganographic scheme based on (7, 4) hamming code for digital images, in *2008 International Symposium on Electronic Commerce and Security, Aug 2008*, (IEEE), pp. 16–21
8. C. Kim, D. Shin, D. Shin, Data hiding in a halftone image using hamming code (15, 11), in *Asian Conference on Intelligent Information and Database Systems, April 2011* (Springer, Berlin, Heidelberg, 2011), pp. 372–381
9. C. Kim, C.N. Yang, *Improving Data Hiding Capacity Based on Hamming Code* (Springer, Netherlands, 2014), pp. 697–706
10. M.H. Mohamed, L.M. Mohamed, High capacity image steganography technique based on LSB substitution method. Appl. Math. Inf. Sci. **10**(1), 259 (2016)
11. B. Jana, D. Giri, S.K. Mondal, Partial reversible data hiding scheme using (7, 4) hamming code. Multimed. Tools Appl. 1–16 (2016)
12. University of Southern California, "The USC-SIPI Image Database", http://sipi.usc.edu/database/database.php
13. C.C. Chang, Y.C. Chou, Using nearest covering codes to embed secret information in grayscale images, in *Proceedings of the 2nd international conference on Ubiquitous information management and communication, Jan 2008* (ACM), pp. 315–320
14. Z. Cao, Z. Yin, H. Hu, X. Gao, L. Wang, High capacity data hiding scheme based on (7, 4) Hamming code. SpringerPlus **5**(1), 175 (2016)

RS (255, 249) Codec Based on All Primitive Polynomials Over GF(2^8)

Jagannath Samanta, Jaydeb Bhaumik, Soma Barman, Sk. G. S. Hossain, Mandira Sahu and Subrata Dutta

Abstract Reed–Solomon (RS) codes are generally employed to detect and correct errors in digital transmission and storage systems. The primitive polynomial has a great role to design any RS codes. In this chapter, a RS (255, 249) codec has been designed and implemented based on sixteen primitive polynomials over GF(2^8) field. The details of theoretical and FPGA synthesis results of the RS (255, 249) codec are presented here. The area in terms of lookup tables and delay of RS (255, 249) codec have been observed for sixteen primitive polynomials. The RS (255, 249) codec based on primitive polynomial, PP3 = $x^8 + x^5 + x^3 + x^2 + 1$, has consumed lowest area compared to all other primitive polynomials. This codec architecture can be employed in M-ary phase-shift keying modulation scheme and ultra-wideband application.

Keywords Galois field · Primitive polynomial · Reed–Solomon code
Phase-shift keying · RiBM algorithm and FPGA

J. Samanta (✉) · J. Bhaumik · M. Sahu · S. Dutta
Haldia Institute of Technology, Haldia, India
e-mail: jagannath19060@gmail.com

J. Bhaumik
e-mail: bhaumik.jaydeb@gmail.com

M. Sahu
e-mail: sahu.mandira@gmail.com

S. Dutta
e-mail: subrataduttaa@gmail.com

S. Barman
Institute of Radio Physics and Electronics, University of Calcutta, Kolkata, India
e-mail: barmanmandal@gmail.com

Sk. G. S. Hossain
Department Of CSE, Aliah University, Newtown, India
e-mail: sarowar25@gmail.com

© Springer Nature Singapore Pte Ltd. 2017
J. Bhaumik et al. (eds.), *Communication, Devices, and Computing*, Lecture Notes in Electrical Engineering 470, https://doi.org/10.1007/978-981-10-8585-7_7

1 Introduction

Reed–Solomon (RS) codes are non-binary cyclic BCH codes. This code has higher error-correcting capability to detect and correct the random errors as well as the burst errors. This code is extensively employed in digital video broadcasting terrestrial (DVB-T) system, satellite and mobile communications, storage or memory systems, etc. [1, 2]. RS codes are encoded and decoded within the general framework of algebraic coding theory which is to map bit streams into abstract polynomials on which a series of mathematical operations are performed. It is required to devise very high-speed implementations of RS decoders due to its huge demand for higher data rates and storage capacity. So, newer and faster designs of the RS decoder are required for further development. Number of decoding algorithms are available in the literatures [2–8]. Among these algorithms, it is very difficult task to choose the best choice due to its number of variables and trade-offs available. Therefore, before making a good choice for a particular application, a thorough research is required.

An uniform comparison is drawn for various algorithms introduced in the literature which helps to select the appropriate architecture for the intended applications. Dual-line architecture of modified Berlekamp–Massey algorithm has been employed for ultra-wideband (UWB) application [9]. The RS (47, 41) decoder is designed and implemented using the extended Euclidean algorithm for intelligent home networking system [10]. An area-efficient RS (23, 17) encoder and decoder circuit is designed and implemented based on local and global optimization algorithm for UWB application [11]. The circuit complexity in terms of two input XOR (XOR2) gate is reduced compared to unoptimized RS codec without affecting its speed [11]. An analytical model and expression for the approximated bit error probability (BEP) for the point-to-point molecular communication systems are proposed using RS codes [12]. The RS (255, 249) codes are designed based on Peterson–Gorenstein–Zierler (PGZ) algorithm to correct up to three-byte errors in WiMedia ultra-wideband technology [13, 14].

In this chapter, RS (255, 249) codec has been designed and implemented based on sixteen primitive polynomials over GF(2^8) field. The rest of this chapter is organized as follows. The M-ary phase-shift keying (MPSK) modulation technique is briefly described in Sect. 2. Different primitive polynomials are presented in Sect. 3. Design of RS (255, 249) codec has been done in Sect. 4. In Sect. 5, theoretical complexity of RS (255, 249) encoder and decoder is provided. FPGA-based simulation and synthesis results of this codec are presented in Sect. 6. Then the chapter is finally concluded in Sect. 7.

2 M-ary Phase-Shift Keying (MPSK) Modulation Technique

In this section, application of RS (255, 249) code in M-ary phase-shift keying (MPSK) modulation scheme is discussed. It is a multi-level modulation technique which permits high data rates within fixed bandwidth constraints. A convenient set of signals for MPSK technique is given in the following:

$$\phi_i(t) = A\cos(\omega_c t + \theta_i), \quad 0 < t \le T_S. \tag{1}$$

where A, ω_c and θ_i are amplitude, angular frequency, and phase shift of the signal. The phase shift θ_i for M-phase angles is shown as follows:

$$\theta_i = 0, 2\pi/M, \ldots, 2(M-1)\pi)/M. \tag{2}$$

In MPSK modulation scheme, data bits select one of M-phase-shifted versions of the carrier to transmit the data. All M-possible waveforms have the same amplitude and frequency but different phases. The signal constellations consist of M equally spaced points on a circle. The RS code is generally employed as the outer code of MPSK modulation scheme in the military communications system. This technique provides better reliability (due to high coding gain) with efficient bandwidth utilization [15, 16]. The MPSK modulation scheme also provides lowest error probability compared to other conventional modulation schemes [17, 18]. The block diagram of RS-coded modulation scheme is shown in Fig. 1. For a systematic RS(N, K) code, a block of information bits is first encoded, where $N \le 2^n$ and n is the number of bits per symbol. The RS-coded bits are passed through an M-ary modulator, and a code symbol comprised of $m = log_2 M$ bits is mapped to one of M signals to be transmitted.

Cellular automata (CA)-based recursive algorithm is introduced for generation of very large primitive polynomial over finite fields [19]. This circuit is employed in a RS-encoded MPSK (RS M-ary phase-shift keying) modulation in Rayleigh fading channel for large code word length. The RS (255, 249) encoder consists of a generator

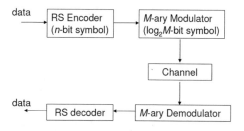

Fig. 1 Block diagram of RS-coded modulation scheme

polynomial block, a field generator block, an n-bit XOR gate block, and a decision control block [19].

3 Primitive Polynomials in GF(2^8) Field

Galois field (GF) or finite field has been employed in many areas such as coding theory, cryptography, number theory, algebraic geometry, polynomial equations, etc [1, 2]. A finite field is completely defined by its primitive polynomial. Primitive polynomials are used in the representation of elements of a finite field. It generates all elements of an extension field from a base field. A polynomial $p(x)$ of degree m over the finite field GF(q) is primitive if it is irreducible and the smallest positive integer n for which $p(x)$ divides $x^n + 1$ is $n = q^m - 1$. A primitive polynomial must have a nonzero constant term; otherwise, it will be divisible by x. The arithmetic operations are also performed using modulo of this primitive polynomial in the GF(p^m) field [20, 21].

If α is a root of a primitive polynomial $p(x)$, then since the order of α is $p^m - 1$ that means that all elements of GF(p^m) can be represented as successive powers of α.

$$GF(p^m) = \{0, 1, \alpha, \alpha^2, \ldots, \alpha^{p^m-2}\}. \tag{3}$$

When these elements are reduced using modulo $p(x)$, they provide the polynomial basis representation of all the elements of the field.

4 Design of RS (255, 249) Encoder and Decoder Block

RS(255, 249) encoder and decoder blocks are discussed in this section.

4.1 RS (255, 249) Encoder Block

RS encoder circuit performs the polynomial division over finite field [22]. The generator polynomial for a triple byte error-correcting RS (255, 249) code is given below:

$$g(x) = (x + \alpha)(x + \alpha^2)(x + \alpha^3)(x + \alpha^4)(x + \alpha^5)(x + \alpha^6) \tag{4}$$

The degree of this generator polynomial is six. Depending on the primitive polynomial, different generator polynomials are generated. The generator polynomial of three-byte RS code is given in Table 1 based on sixteen different primitive polynomials (PP1, ..., PP16) in GF(2^8) field.

These primitive polynomials are important to design any encoder and decoder circuit. Hansen–Mullen's conjecture on the distributions of primitive polynomials over finite field is discussed in [21] by considering second coefficient and odd or even characteristic.

Table 1 Generator polynomials of three-byte error-correcting RS code in GF(2^8)

Name	Primitive polynomial	Decimal equivalent	Generator polynomial
PP1	$x^8 + x^4 + x^3 + x^2 + 1$	285	$x^6 + 126x^5 + 4x^4 + 158x^3 + 58x^2 + 49x + 117$
PP2	$x^8 + x^5 + x^3 + x + 1$	299	$x^6 + 126x^5 + 71x^4 + 171x^3 + 26x^2 + 168x + 50$
PP3	$x^8 + x^5 + x^3 + x^2 + 1$	301	$x^6 + 126x^5 + 105x^4 + 119x^3 + 135x^2 + 21x + 219$
PP4	$x^8 + x^6 + x^3 + x^2 + 1$	333	$x^6 + 126x^5 + 19x^4 + 58x^3 + 63x^2 + 188x + 245$
PP5	$x^8 + x^6 + x^4 + x^3 + x^2 + x + 1$	351	$x^6 + 126x^5 + 253x^4 + 90x^3 + 30x^2 + 5x + 76$
PP6	$x^8 + x^6 + x^5 + x + 1$	355	$x^6 + 126x^5 + 202x^4 + 225x^3 + 218x^2 + 194x + 237$
PP7	$x^8 + x^6 + x^5 + x^2 + 1$	357	$x^6 + 126x^5 + 238x^4 + 41x^3 + 42x^2 + 119x + 110$
PP8	$x^8 + x^6 + x^5 + x^3 + 1$	361	$x^6 + 126x^5 + 166x^4 + 248x^3 + 218x^2 + 148x + 96$
PP9	$x^8 + x^6 + x^5 + x^4 + 1$	369	$x^6 + 126x^5 + 54x^4 + 163x^3 + 190x^2 + 75x + 26$
PP10	$x^8 + x^7 + x^2 + x + 1$	391	$x^6 + 126x^5 + 160x^4 + 34x^3 + 167x^2 + 228x + 95$
PP11	$x^8 + x^7 + x^3 + x^2 + 1$	397	$x^6 + 126x^5 + 240x^4 + 144x^3 + 138x^2 + 196x + 25$
PP12	$x^8 + x^7 + x^5 + x^3 + 1$	425	$x^6 + 126x^5 + 121x^4 + 5x^3 + 183x^2 + 173x + 82$
PP13	$x^8 + x^7 + x^6 + x + 1$	451	$x^6 + 126x^5 + 197x^4 + 44x^3 + 104x^2 + 120x + 102$
PP14	$x^8 + x^7 + x^6 + x^3 + x^2 + x + 1$	463	$x^6 + 126x^5 + 177x^4 + 160x^3 + 40x^2 + 196x + 184$
PP15	$x^8 + x^7 + x^6 + x^5 + x^2 + x + 1$	487	$x^6 + 126x^5 + 110x^4 + 232x^3 + 138x^2 + 51x + 152$
PP16	$x^8 + x^7 + x^6 + x^5 + x^4 + x^2 + 1$	501	$x^6 + 126x^5 + 218x^4 + 123x^3 + 184x^2 + 230x + 203$

The transmitted code word polynomial $c(x)$ can be expressed in terms of input message $m(x)$, generator polynomial $g(x)$, remainder polynomial $r(x)$, and quotient polynomial $q(x)$.

$$c(x) = x^{n-k}m(x) + [x^{n-k}m(x) \bmod g(x)] = x^{n-k}m(x) + r(x) = q(x)g(x) \quad (5)$$

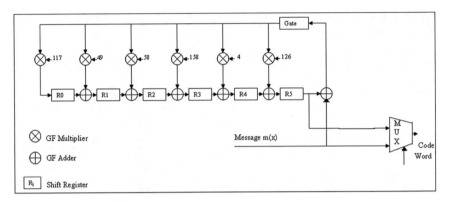

Fig. 2 Architecture of RS (255, 249) encoder

Figure 2 shows the architecture of a linear-feedback shift register (LFSR) systematic RS (255, 249) encoder which consists of six GF constant multipliers, six GF adders, and six shift registers. During the first 249 clock cycles, 'Gate' is closed and 'MUX' will pass the simply message symbol through code word output. During the last 6 clock cycles, 'Gate' is opened then parity symbols are generated and passed through the 'MUX.' The RS (255, 249) decoder is described in the following subsection.

4.2   RS (255, 249) Decoder Block

The RS algebraic or hard decoding procedures can correct both errors and erasures. Figure 3 shows the general block diagram of the RS decoder. The RS decoder has main five blocks which are discussed in this section.

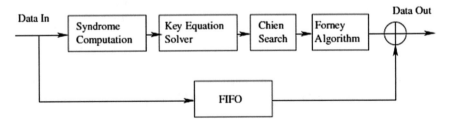

Fig. 3 Block diagram of RS decoder architecture

(a) Syndrome Calculator: Syndrome computation is performed to detect if the code word contains any errors or not [23]. For a t error-correcting code, the $2t$ syndromes are computed using following equation:

$$S_j = R(\alpha^j) = \sum_{i=0}^{n-1} R_i(\alpha^j)^i; \quad 1 \le j \le 2t$$

$$S_j = c(\alpha^j) + e(\alpha^j) = e(\alpha^j) \tag{6}$$

where α is a root of the generator polynomial $g(x)$. If all syndromes are not zero, then the received code word is corrupted. Syndromes are used to derive error locator and error evaluator polynomials. The syndromes are calculated by substituting the $2t$ roots of the generator polynomial $g(x)$ into $r(x)$.

(b) Key Equation Solver Block: KES block computes the error locator polynomial $\Lambda(x)$ and error magnitude polynomial $\Omega(x)$ from the syndrome polynomial $S(x)$. The key equation can be represented as

$$S(x)\Lambda(x) = \Omega(x) \bmod x^{2t}. \tag{7}$$

In practice, any one algorithm, namely Euclidean algorithm, modified Euclidean algorithm [8], simplified degree computationless modified Euclid's algorithm (S-DCME) [3], Berlekamp–Massey algorithm [2, 5], or inversionless Berlekamp–Massey (iBM) algorithm [4, 6, 7], is used to solve the key equation solver (KES) block. The decoding process which finds the error locator and error evaluator polynomial involves high computational complexity. These also affect the speed and the hardware complexity of the RS decoders.

In 2001, Sarwate et al. [7] proposed a decoding algorithm based on iBM which is popularly known as 'Reformulated iBM algorithm' (RiBM). This algorithm provides a regular semi-systolic architecture with minimum critical path delay due to a finite field multiplier and an adder. Irregular structure and longer delay of the BM architecture have been reduced by the reformulated inversionless Berlekamp–Massey (RiBM) architecture [7]. This RiBM algorithm [7] removes the unnecessary components of the KES block and makes the architecture lower hardware complexity and shorter latency than conventional BM architectures. It is a regular decoder architecture and requires only $2t$ clock cycles. In 2010, Park et al. [24] modified the RiBM algorithm by considering the pipelining and folding concept which is known as 'Pipelined Reformulated Inversionless Berlekamp–Massey (PRiBM) algorithm.'

Reformulated inversionless and dual-line implementation of the modified Berlekamp–Massey have the smallest critical path delay among all other alternatives of the KES blocks.

(c) **Error Locator**: The RS decoder now evaluates the root of $\Lambda(x)$ polynomial to find the position of errors using the Chien search algorithm, where $\Lambda(x)$ is the error locator polynomial.

(d) **Magnitude Finder**: Error magnitude is evaluated by the Forney algorithm. Equation for computing the error value is as follows:

$$Y_l = \frac{\Omega(x)}{\Lambda'(x)} \quad for \ x = \alpha^{i_l.} \tag{8}$$

where $\Lambda'(x)$ is the derivative of $\Lambda(x)$.

The fifth block is error correction block which is simply XORing between received code word with calculated error pattern. The detail theoretical analysis of RS (255, 249) is presented in the following section.

5 Theoretical Complexity of RS (255, 249) Encoder and Decoder

Three-byte RS (255, 249) codec has been designed based on sixteen primitive polynomials in GF(2^8) field. The details of theoretical complexity of codec have been given in Table 2. Theoretical complexity is given in terms of two input XOR (XOR2) gate, multiplexer (MUX), and flip-flop (FF). From the Table 2, it is observed that maximum area complexity of RS (255, 249) decoder is KES block.

In the next section, RS (255, 249) codec is simulated and synthesized using the electronic design automation tool.

6 Synthesis Results

All the blocks are represented in Verilog HDL language. All designs are simulated and synthesized using Xilinx-based vertex 4 FPGA device family (target device:4vfx12sf363-12). For input sequences [48, 75, 252, 212, 212, 247, 255, 184, 12, ... 54, 54, 54], corresponding six parity bytes are [35, 17, 47, 181, 241, 125] which is shown in Fig. 4.

Table 2 Theoretical complexity in terms of XOR2 and AND2 gates of RS (255, 249) encoder and decoder block

Block		PP1	PP2	PP3	PP4	PP5	PP6	PP7	PP8	PP9	PP10	PP11	PP12	PP13	PP14	PP15	PP16
Enc	XOR2	171	207	208	207	187	206	211	201	202	201	186	203	201	202	199	188
	MUX	1	1	1	1	1	1	1	1	1	1	1	1	1	1	1	1
Synd	XOR2	72	77	79	79	75	77	79	75	77	75	76	76	75	77	76	76
	FF	48	48	48	48	48	48	48	48	48	48	48	48	48	48	48	48
KES ×10³	XOR2	14.6	15.6	13.7	15.6	15.8	16	15.4	16.6	14.9	15.1	15.4	17.7	14.8	15.9	15.1	17.0
	AND2	15.0	16.0	14.1	16.0	16.2	16.4	15.8	17.0	15.3	15.5	15.8	18.1	15.2	16.3	15.5	17.4
Chien	XOR2	51	53	56	55	53	54	55	55	54	52	52	52	53	54	52	55
	MUX	4	4	4	4	4	4	4	4	4	4	4	4	4	4	4	4
	FF	32	32	32	32	32	32	32	32	32	32	32	32	32	32	32	32
Forney	XOR2	321	351	340	359	345	355	357	353	344	338	345	368	335	354	342	361
	AND2	238	250	234	252	252	255	250	262	244	244	247	270	242	254	244	266
	MUX	4	4	4	4	4	4	4	4	4	4	4	4	4	4	4	4
	FF	32	32	32	32	32	32	32	32	32	32	32	32	32	32	32	32
Total ×10³	XOR2	15.1	16.1	14.2	16.1	16.3	16.5	15.9	17.1	15.4	15.6	15.9	18.3	15.3	16.4	15.6	17.5
	AND2	15.2	16.2	14.3	16.2	16.4	16.6	16.0	17.2	15.5	15.7	16.0	18.3	15.4	16.5	15.7	17.6
	MUX	9	9	9	9	9	9	9	9	9	9	9	9	9	9	9	9
	FF	112	112	112	112	112	112	112	112	112	112	112	112	112	112	112	112

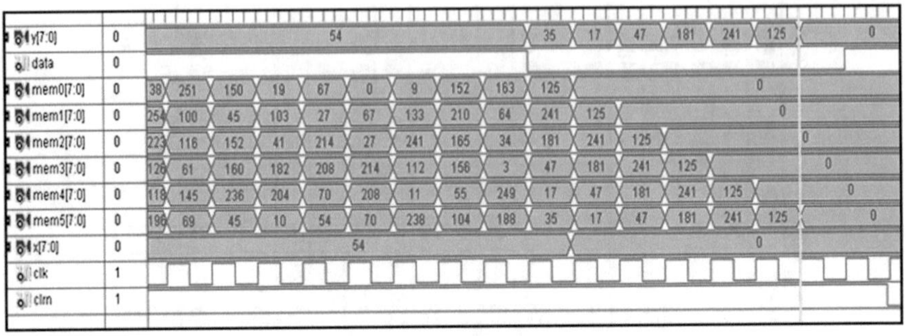

Fig. 4 Simulation waveform of RS (255, 249) encoder for PP1

The FPGA-based synthesis results of different blocks of the RS (255, 249) codec for sixteen primitive polynomials are presented. In this chapter, KES block has been implemented using RiBM algorithm. The area (in lookup table (LUT)), maximum combinational path delay (in ns), and operating frequency (in MHz) are shown in the Table 3; it is observed that the maximum area is consumed in PP12. FPGA synthesis results of the RS (255, 249) encoder and decoder block are presented in Table 3.

From Table 3, it is observed that the maximum area is consumed in PP12 and minimum is in PP3. Delay information is also observed from Table 3.

Fig. 5 shows the area comparison in terms of LUTs for sixteen different RS (255, 249) codec architecture. It is observed that minimum and maximum area is consumed in PP3 and PP12, respectively.

Table 3 FPGA synthesis results of RS (255, 249) encoder and decoder block

Block		PP1	PP2	PP3	PP4	PP5	PP6	PP7	PP8	PP9	PP10	PP11	PP12	PP13	PP14	PP15	PP16
Enc	Slice	81	86	87	86	78	85	87	97	85	84	88	85	88	96	85	79
	LUT	154	166	167	166	150	165	168	189	165	163	171	165	169	186	164	153
	Delay	4.98	5.02	4.98	5.14	4.98	5.06	4.98	5.14	5.02	4.98	5.02	5.06	5.02	5.02	5.07	4.98
	Freq.	358	354	346	348	357	354	355	369	363	357	372	349	311	362	310	376
Synd	Slice	77	77	77	78	78	78	78	79	78	74	74	80	76	75	75	77
	LUT	143	141	142	144	144	144	143	146	145	136	136	149	139	138	138	142
	Delay	–	–	–	–	–	–	–	–	–	–	–	–	–	–	–	–
	Freq.	443	444	444	444	444	444	443	443	444	444	444	444	444	444	444	444
KES	Slice	2073	2111	1955	2170	2064	2215	2170	2198	2105	2111	2257	2334	2222	2000	2139	2166
	LUT	3609	3677	3425	3779	3592	3869	3775	3829	3660	3681	3936	4020	3872	3483	3724	3770
	Delay	31.6	30.5	33.3	38.1	35.0	30.1	31.7	33.8	29.6	30.7	28.7	31.2	33.1	32.1	31.8	34.3
	Freq.	–	–	–	–	–	–	–	–	–	–	–	–	–	–	–	–
Chien search	Slice	79	81	80	80	84	81	80	80	79	79	81	81	78	81	81	83
	LUT	142	146	144	144	150	145	144	144	142	142	145	146	141	146	146	148
	Delay	7.81	7.81	7.81	7.89	7.89	7.89	7.89	7.89	7.89	7.89	7.89	7.89	7.89	7.89	7.89	7.89
	Freq.	460	460	460	460	460	460	460	460	460	460	460	460	460	460	460	460
Forney	Slice	133	136	135	137	138	135	134	127	133	132	137	135	134	138	137	142
	LUT	235	240	238	241	244	239	237	241	235	234	242	239	238	244	242	250
	Delay	14.5	14.3	14.2	13.8	15.2	14.1	14.3	13.9	14.4	14.2	14.3	14.1	13.9	14.1	14.7	14.5
	Freq.	469	463	469	469	469	469	469	467	469	469	469	469	469	469	469	469
Total	Slice	2443	2491	2334	2551	2442	2594	2549	2591	2480	2480	2637	2715	2598	2390	2517	2547
	LUT	4283	4370	4116	4474	4280	4562	4467	4549	4347	4356	4630	4719	4559	4197	4414	4463

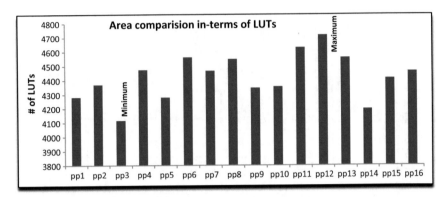

Fig. 5 Area comparison in terms of LUTs

7 Conclusion

In this chapter, RS (255, 249) codec has been designed and implemented for sixteen different primitive polynomials in GF(2^8) field. The area and delay information of different blocks of RS (255, 249) code are presented here. Minimum and maximum hardware complexity are found for PP3 and PP12, respectively. This codec architecture can be used in M-ary PSK modulation technique. This will help the designers for further development of VLSI based on RS codec architectures for different resource constraint applications.

References

1. S. Lin, D.J. Costello, *Error Control Coding: Fundamentals and Applications*. Prentice-Hall (1983)
2. S.B. Wicker, V.K. Bhargava, *Reed Solomon Codes and Their Applications*. IEEE Press (1994)
3. J. Baek, M.H. Sunwoo, Low hardware complexity key equation solver chip for ReedSolomon decoders, in *Proceedings of IEEE Asian Solid-State Circuits Conference* (Jeju, Korea, 2007), p. 5154
4. I.S. Reed, M.T. Shih, T.K. Truong, VLSI design of inverse-free Berlekamp Massey algorithm, in *Proceedings of the Institution of Electrical Engineers*, vol. 138 (1991), pp. 295–298
5. H.C. Chang, C.B. Shung, New serial architecture for the Berlekamp Massey algorithm. IEEE Trans. Commun. **47**(4), 441–443 (1999)
6. J.H. Jeng, T.K. Truong, On decoding of both errors and erasures of a Reed-Solomon code using an inverse-free Berlekamp-Massey algorithm. IEEE Trans. Commun. **47**(10), 1488–1494 (1999)
7. D.V. Sarwate, N.R. Shanbhag, High-speed architectures for Reed-Solomon decoder. IEEE Trans. VLSI Syst. **9**(5), 641–655 (2001)
8. Y.W. Chang, T.K. Truong, J.H. Jeng, VLSI architecture of modified Euclidean algorithm for reed-Solomon code. Inform. Sci. **155**(1), 139–150 (2003)

9. A. Kumar, S. Sawitzki, High Throughput and Low power Reed Solomon Decoder for Ultra Wide Band, in *Proceedings of Intelligent Algorithms in Ambient and Biomedical Computing*, vol. 7 (2006), pp. 299–316
10. I.S. Jin, Design and implementation of efficient Reed-Solomon decoder for intelligent home networking, in *Proceedings of Conference (FGCN), CCIS 265* (2011), pp. 261–268
11. J. Samanta, J. Bhaumik, S. Barman, FPGA based area efficient RS (23, 17) codec. Microsyst. Technol. **23**(3), 639–650 (2017)
12. M.B. Dissanayake, Y. Deng, A. Nallanathan, E.M.N. Ekanayake, M. Elkashlan, Reed Solomon codes for molecular communication with full absorption receiver. IEEE Commun. Lett. (2017)
13. (255, 249) Reed Solomon Decoder, Algorithm data sheet. Blue Rum Consulting Limited (2009)
14. G. Heidari, *WiMedia UWB: Technology of Choice for Wireless USB and Bluetooth.* Wiley (2008). ISBN: 978-0-470-51834-2
15. R.D. Cideciyan, E. Eleftheriou, Concatenated Reed-Solomon/convolutional coding scheme for data transmission in CDMA cellular systems, in *IEEE Vehicular Technology Conference* (1994), pp. 1369–1373
16. Kar Peo Yar, Design and analysis of short packet and concatenated coded communication systems, Ph.D. Thesis, The University of Michigan, 2007
17. S.N. Ramlan, R. Mohamad, N. Arbain, Implementation of M-ary Phase Shift Keying (PSK) base band modem on Texas instrument digital signal processor TMS320C6713, in *Proceedings in IEEE International Conference on Computer Applications and Industrial Electronics (ICCAIE)* (2011), pp. 627–632
18. M. Wan, N. Zhang, Searching IP Blocks: Application Specific Components, https://people.eecs.berkeley.edu/~newton/Classes/EE290sp99/pages/hw2/asc.htm
19. D. Bhattacharya, D. Mukhopadhyay, D. Roy, Chowdhury, A cellular automata based approach for generation of large primitive polynomial and its application to RS-Coded MPSK modulation. Lecture Notes in Computer Science vol. 4173 (2006), pp. 204–214
20. T. Hansen, G.L. Mullen, Primitive polynomials over finite fields. Math. Comput. **59**(200), 639–643 (1992)
21. W. Han, The distribution of coefficient of primitive polynomials over finite fields. Cryptogr. Comput. Number Theory **20**, 43–57 (2001)
22. C. Xiaojun, G. Jun, L. Zhihui, RS encoder design based on FPGA, in *Proceedings of 2nd IEEE, ICACC 2010*, vol. 1 (2010), pp. 419-421
23. R. Huynh, G.E. Ning, Y. Huazhong, A low power error detection in the syndrome calculator block for Reed-Solomon codes: RS(204, 188). J. Tsinghua Sci. Technol. **14**(4), 474–477 (2009)
24. J.I. Park, K. Lee, C.S. Choi, H. Lee, High-speed low-complexity Reed-Solomon decoder using pipelined Berlekamp-Massey algorithm and its folded architecture. J. Semicond. Technol. Sci. **10**(3), 193–202 (2010)

Secure User Authentication System Using Image-Based OTP and Randomize Numeric OTP Based on User Unique Biometric Image and Digit Repositioning Scheme

Ramkrishna Das, Sarbajit Manna and Saurabh Dutta

Abstract Proposed system introduces a combined one time password (OTP)-based authentication system where image OTP is used in first level and numeric OTP is used in the second level of authentication. Server randomly selects an image as image OTP and encrypts it using user unique biometric image and user-defined Bit-wise Masking and Alternate Sequence (BWMAS) operation. This encrypted image OTP will be shared and to be decrypt by the user. Then the system generates a large random number as first part of numeric OTP within a range where range value is derived from the user password. Second part of numeric OTP is generated from the values of randomly selected blocks of the randomly selected pixels of user biometric image. Finally we combine those OTPs using alternate merging and generate intermediate numeric OTP which will be shared from server to user. Final numeric OTP will be generated in user and server end from intermediate OTP using user-defined digit repositioning scheme chosen by user. Random generation of numeric and image OTP, distribution of encrypted image OTP and formation of final numeric OTP using digit repositioning scheme impose a great security to the system.

Keywords Image-based OTP · Numeric OTP · BWMAS operation · Random selection · Biometric image · Digit repositioning scheme

R. Das (✉)
Department of Computer Applications, Haldia Institute of Technology, Haldia, West Bengal, India
e-mail: ramkrishnadas9@gmail.com

S. Manna
Department of Computer Science, Ramakrishna Mission Vidyamandira, Belur Math, Howrah 711202, West Bengal, India
e-mail: sarbajitonline@gmail.com

S. Dutta
Department of Computer Applications, Dr. B. C. Roy Engineering College, Durgapur 713206, West Bengal, India
e-mail: saurabhdutta06061973@gmail.com

© Springer Nature Singapore Pte Ltd. 2017
J. Bhaumik et al. (eds.), *Communication, Devices, and Computing*, Lecture Notes in Electrical Engineering 470, https://doi.org/10.1007/978-981-10-8585-7_8

1 Introduction

Traditional numeric- and image-based OTP is not so much secured as distribution of OTP is done through public communication channel [1–3]. So we have proposed a combined OTP system where encrypted random image-based OTP followed by a randomized two-phase numeric OTP is used for authentication. Thus, increase the security of the OTP-based authentication. In this paper, Sect. 2 and Sect. 3 discuss the background study and preliminaries, respectively. Section 4 describes the overall procedure. Section 5, Sect. 6 and Sect. 7 represent formation of OTP at server, distribution of OTP, extraction of OTP at user end and authentication, respectively. Experimental results are described in Sect. 8. Section 9 shows the comparison with existing OTP system, and Sect. 10 draws conclusions.

2 Background Study

Srinivas and Janaki proposed an approach where random numbers are generated from extracted features of image, used as OTP which forms a strong factor for authentication [4]. Huang et al. propose an OTP method that generates a unique passcode based on both time stamps and sequence numbers [5]. Shesashaayee and Sumathy proposed a scheme that combines the secret PIN and OTP where PIN is used to encrypt the OTP [6]. Vishwakarma and Gangrade proposed an approach that system uses random image and text-based OTP generation with SHA-512 algorithm and again encryption by using ECC to develop OTP [7]. Kushwaha proposed a scheme where OTP will be generated by combining the image and numeric value [8]. All those OTP systems are not so secured as distribution of OTP is done through public channel. Proposed system tries to overcome those security issues.

3 Preliminaries

3.1 BWMAS Operation (Bit-wise Masking and Alternate Sequence)

BWMAS will be performed between the binary representation of original image and key image. The procedure will follow the following rules:

Rule 1: For even bit position of original image

If corresponding bit position value of enveloped image $= 0$, then output is corresponding bit value of original image. If corresponding bit position value of enveloped image $= 1$, then output is complement of corresponding bit value of original image.

Rule 2: For odd bit position of original image

If corresponding bit position value of enveloped image $= 1$, then output is corresponding bit value of original image. If corresponding bit position value of enveloped image $= 0$, then output is complement of corresponding bit value of original image.

4 Overall Procedure

Step I: User provides information of username, password and biometric unique image to the authentication server and makes the registration process. User makes login attempt into the system, and if user provides proper username and password, then first-level authentication will be done and will go for next.

Step II: Server generates image OTP by randomly selecting an image from its image database. Server encrypts OTP image by user biometric image using Bit-wise Masking and Alternate Sequence (BWMAS) operation and generates encrypted OTP image.

Step III: Server generates first-level numeric OTP by generating a random number into the range from 0 to a range value which is defined from the ASCII values of the characters from the password of the user. Server generates second-level numeric OTP by fetching the values from the randomly selected blocks of the randomly selected pixels from the biometric image of the user. Both the OTPs are alternately merged with each other and generate the intermediate OTP and server executes user-defined OTP digit repositioning algorithm chosen by the user and generate the final numeric OTP.

Step IV: Intermediate OTPs are divided into two parts and sent to user by email and message. Encrypted OTP image will be separately sent to user from server by email.

Step V: User extracts image OTP from the encrypted OTP image by using BWMAS operation with user's biometric image. User also generates final numeric OTP from the intermediate OTP by using digit repositioning algorithm chosen by the user at registration time. Image OTP and the final numeric OTP are used for authentication.

5 Formation of OTP at Authentication Server End

User provides username, password, unique biometric image and choice for digit repositioning scheme to server as input.

5.1 Algorithm for Generation and Encryption of Image OTP

Step I: Server randomly selects an image as image OTP, and its width (w) and height (h) are calculated. User biometric images' width (wk) and height (hk) are calculated, where wk <= w and hk <= h. Two arrays IMG_RGB and IMG_KEY of size w * h * 24 store bit values of 24-bit RGB part of OTP image and biometric image, respectively.

Step II: BWMAS operation is performed between IMG_KEY array and IMG_RGB array, and the result is stored in IMG_CON array. Encrypted image OTP is generated from IMG_CON [w * h] and sent to user.

5.2 Algorithm for Generating Numeric OTP

Step I: Generate the ASCII values from each character of the user password and multiply each of the ASCII values which generate the range value. Generate a long random value in between the limits of zero and range value derived earlier and it will be used as a first-level numeric OTP. Randomly select 4 pixels and 4 block numbers from the user biometric image. Fetch the bit values from those randomly selected blocks of randomly selected pixels and convert them into decimal values. Thus, generate second-level numeric OTP.

Step II: Perform alternate merging between each digit of first- and second-level numeric OTP. Thus, generate the intermediate numeric OTP which will be shared to user. Server executes one of the digit repositioning schemes chosen by user and generates the final OTP, stores and uses it for user authentication in future.

5.3 Algorithm for Digit Repositioning Schemes

Positional reverse prime normal non prime (PRPNNP)
Store each digits in an array pro[]. Fetch and reverse the prime position digit's value and store them in array pro_f[]. Non prime position's digits are stored to array pro_f[] without any changes.
Positional reverse non prime normal prime (PRNPNP)
Store each digits value in an array pre[]. Fetch and reverse the non prime position digit's value and store them in array pre_f[]. Prime position digits are stored to array pre_f[] without any changes.
Continuous reverse prime normal non prime (CRPNNP)

Store each digits value in an array cre[]. Fetch and reverse the prime position digit's value and store them in array cre_f[] continuously. Non prime position digits are stored to array cre_f[] continuously without any changes.

Continuous reverse non prime normal prime (CRNPNP)

Store each digits value in an array cro[]. Fetch and reverse the non prime position digit's values and store them in array cro_f[] continuously. Prime position digits are stored to array cro_f[] continuously without any changes.

6 Distribution of Image-Based OTP and Numeric OTP

Encrypted image OTP and intermediate OTP are divided into two parts, and server sends these OTPs to user by email and message. Final OTP has not been shared between server and user that have to be generated from intermediate OTP separately in the server end and user end.

7 Extraction of OTP at User End and Authentication

User performs BWMAS operation between user biometric image and encrypted image OTP and generates original OTP image. User executes digit repositioning algorithm chosen by user at registration time on intermediate OTP and generates final OTP. Theses image OTP and final numeric OTP will be used for authentication.

8 Results and Comparisons

8.1 Inputs at User Registration Time to Authentication System

User provides user id, password, unique biometric image and choice of digit repositioning algorithm to the server. Figures 1 and 2 show the activities.

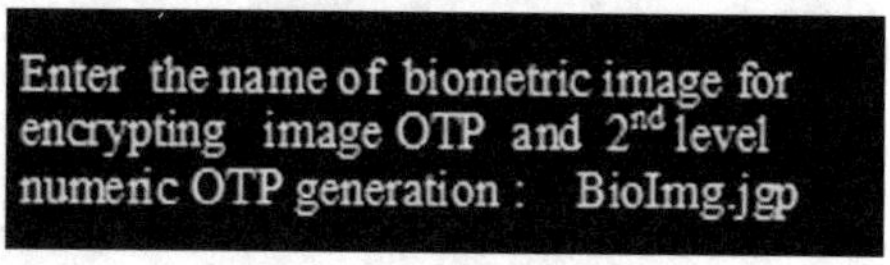

Fig. 1 Input taking for biometric image

Enter the name of biometric image for encrypting image OTP and 2nd level numeric OTP generation :   BioImg.jgp

Fig. 2 User biometric image

8.2 Formation of Image-Based OTP at Server End

Server will randomly select a image (here M.jpg) as image OTP which will be encrypted by user biometric image and shared to user as encrypted image OTP (Fig. 3).

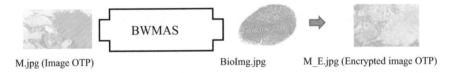

M.jpg (Image OTP)   BioImg.jpg   M_E.jpg (Encrypted image OTP)

Fig. 3 Encryption of image OTP using user biometric image at server end

8.3 Generation of First-Level Numeric OTP

See Fig. 4.

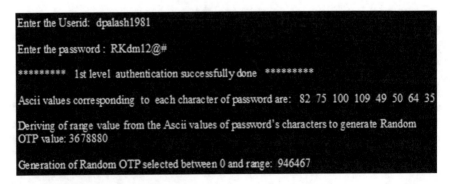

Fig. 4 Generation of first-level numeric OTP

8.4 Generation of Biometric Image-Based Second-Level Numeric OTP

See Fig. 5.

```
Four number of pixels values selected randomly from biometric image:
0, 2887, 6078, 9773

Four number of blocks values selected randomly applied for corresponding pixels to fetch
binary values from biometric image:
0 1 0 3

Decimal value generated from binary values fetched from random block position of randomly
selected pixels:
167720366

Digits of the second level numeric OTP:
1 6 7 7 2 0 3 6 6
```

Fig. 5 Generation of second-level numeric OTP

8.5 Intermediate OTP Generated After Alternate Merge Between Two OTPs

See Fig. 6.

Fig. 6 Intermediate OTP

```
9 1 4 6 6 7 4 7 6 2 7 0 3 6 6
```

8.6 Generation of Final OTP

Options for OTP repositioning algorithms chosen by the user and generation of final numeric OTP generated by server.

See Figs. 7 and 8.

> 1. Positional Reverse Prime Normal Non Prime
> 2. Positional Reverse Non Prime Normal Prime
> 3. Continuous Reverse Prime Normal Non Prime
> 4. Continuous Reverse Non Prime Normal Prime
>   Enter Your Choice   1

Fig. 7 Option for OTP digit repositioning scheme

> 3 7 4 6 6 7 4 7 6 2 1 0 9 6 6

Fig. 8 Final numeric OTP value

8.7 Distribution of Encrypted Image OTP and Intermediate Numeric OTP Through User Email and Message

See Fig. 9.

Fig. 9 Distribution of image OTP and numeric OTP through email and SMS

8.8 Extraction of Image OTP and Numeric OTP at User End

See Figs. 10 and 11.

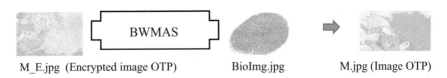

M_E.jpg (Encrypted image OTP)   BioImg.jpg   M.jpg (Image OTP)

Fig. 10 Generation of image OTP

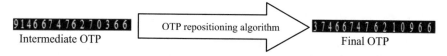

Fig. 11 Generation of numeric final OTP

8.9 User Authentication

See Fig. 12.

Fig. 12 Login attempt by unauthenticated and authenticate user

9 Comparison of Existing OTP Systems and Security Analysis

See Table 1.

Table 1 Comparison with existing one time password (OTP) systems

Proposer of the system	Core idea
Srinivas and Janaki [4]	OTP based on extracted features of image
Huang et al. [5]	OTP based on time stamps and sequence numbers
Vishwakarma and Gangrade [7]	Random image- and text-based OTP generation with SHA-512 algorithm
Proposed system	Encrypted random image OTP and two-phase numeric OTP generated from user biometric image

If someone hacks intermediate OTP and biometric image, still the system is secured as digit repositioning scheme and BWMAS operation are kept secret. Table 2

Table 2 Security analysis of the proposed system

Total number of digits in intermediate numeric OTP	Number of attempts needed for all possible combination of digits of OTP	Size of biometric image in bits (w(width) * h(height) * 32)	Number of executions needed for all possible combination of bits from biometric image to generate second-level numeric OTP
15	Factorial (15)	100 * 200 * 32	Factorial (640000)/factorial (32) * factorial (640000 − 32)
20	Factorial (20)	100 * 300 * 32	Factorial (960000)/factorial (32) * factorial (960000 − 32)
25	Factorial (25)	100 * 400 * 32	Factorial (1280000)/factorial (32) * factorial (1280000 − 32)

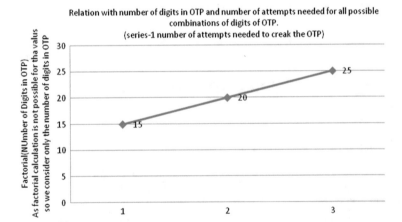

Fig. 13 Relationship between numbers of digits in OTP and number of attempts needed for all possible combinations of digits of OTP

shows security analysis of the system, and Fig. 13 shows the relationship between number of digits in OTP and number of attempts to creak it.

10 Conclusions

Five levels of securities are present in the proposed system: user id- and password-based authentication, image OTP-based authentication, distribution of image OTP by encrypting it using biometric image, two level of numeric OTP generated from biometric image and ASCII values of text. Random selection of image OTP and distribution of encrypted image OTP provide more security; as if the OTP is hacked, still the OTP may not be retrieved without user biometric unique image. First-level numeric OTP and the pixels and blocks of biometric image are randomly selected for the second-level numeric OTP. Then the digits of the both OTP are alternately merged. So no random guessing can retrieve the numeric OTP. Intermediate numeric OTP is distributed to user. If this OTP is hacked, still the system is totally secured as both the server and user generate final OTP from intermediate OTP using digit repositioning scheme. Thus, the security is enhanced to a great extent.

References

1. V.R.S. Mali, Graphical password as an OTP. Int. J. Eng. Comput. Sci. **6**(1), 20090–20095 (2017)
2. https://en.wikipedia.org/wiki/One-time_password
3. https://www.bobcards.com/otp-procedure.htm
4. K. Srinivas, V. Janaki, A novel approach for generation of OTP'S using image's, in *International Conference on Computational Modeling and Security (CMS), 2016, Procedia Computer Science*, vol. 85 (ScienceDirect, 2016), pp. 511–518, https://doi.org/10.1016/j.procs.2016.05.206
5. Y. Huang, Z. Huang, H. Zhao, X. Lai, A new one-time password method, in *International Conference on Electronic Engineering and Computer Science, 2013, IERI Procedia*, vol. 4 (ScienceDirect, 2013), pp. 32–37, https://doi.org/10.1016/j.ieri.2013.11.006
6. A. Shesashaayee, D. Sumathy, OTP encryption techniques in mobiles for authentication and transaction security. Int. J. Innov. Res. Comput. Commun. Eng. **2**(10), 6192–6201 (2014)
7. N. Vishwakarma, K. Gangrade, Secure image based one time password. Int. J. Sci. Res. (IJSR) **5**(11), 680–683 (2016)
8. B.K. Kushwaha, An approach for user authentication one time passward (Numeric and graphical) scheme. J. Glob. Res. Comput. Sci. **3**(11), 54–57 (2012)

Application of RCGA in Optimization of Return Loss of a Monopole Antenna with Sierpinski Fractal Geometry

Ankan Bhattacharya, Bappadittya Roy, Shashibhushan Vinit
and Anup K. Bhattacharjee

Abstract This paper presents a monopole antenna design using Sierpinski triangle fractal geometry with an improved frequency response. The antenna response has been optimized using the genetic algorithmic approach. Triangular slots have been incorporated in the ground plane section, which has an effect on antenna return loss. A triangular shaped patch with Sierpinski triangle geometry has been placed above Beryllia (99.5%) substrate having an electrical permittivity of 6.5. The resonant frequency peak of proposed antenna is exactly at 2.45 GHz with an impedance bandwidth of 650 MHz. The antenna finds its application in 2.4 GHz (2.41–2.48 GHz) WLAN band with a maximum realized gain of 3.16 dBi at the resonating frequency.

Keywords Microstrip patch antenna · Sierpinski triangle · Genetic algorithm

1 Introduction

Monopole antenna with fractal geometry is gaining huge popularity in the field of wireless communication. The most common disadvantage of the conventional Microstrip Patch Antenna (MPA) is narrow impedance bandwidth and significant return loss. High impedance bandwidth can be achieved by the implementation of various fractal structures, modifying the patch or the ground plane, without changing the substrate. Triangular, rectangular, square, trapezoidal, circular, elliptical, etc., are few popular slot geometries that can be used together with a tuning stub for obtaining a wider bandwidth, as found in the articles [1–6]. In this article, a monopole antenna has been designed using the popular 'Sierpinski triangle' fractal geometry with an improved frequency response. For optimization of the antenna frequency response,

A. Bhattacharya (✉) · B. Roy · S. Vinit · A. K. Bhattacharjee
Department of Electronics and Communication Engineering,
National Institute of Technology, Durgapur, India
e-mail: bhattacharya.ankan@ieee.org

A. Bhattacharya
Department of Electronics and Communication Engineering,
Mallabhum Institute of Technology, Bishnupur, India

© Springer Nature Singapore Pte Ltd. 2017
J. Bhaumik et al. (eds.), *Communication, Devices, and Computing*, Lecture Notes
in Electrical Engineering 470, https://doi.org/10.1007/978-981-10-8585-7_9

Fig. 1 Sierpinski triangular
monopole antenna

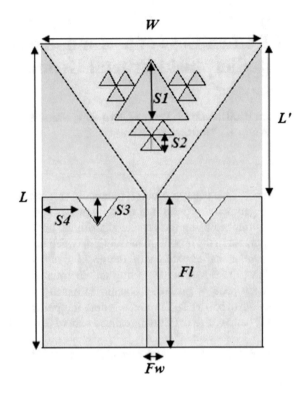

real-coded genetic algorithm (RCGA) has been implemented. The ground plane has
been modified with triangular slots, which have an effect on antenna return loss. A
triangular shaped patch with Sierpinski triangle geometry is placed on top of Beryllia
(99.5%) substrate having an electrical permittivity of 6.5. The proposed antenna
resonates exactly at 2.45 GHz with an impedance bandwidth of almost 650 MHz.
The monopole antenna finds its application in 2.4 GHz (2.41–2.48 GHz) WLAN
band with a maximum realized gain of 3.362 dBi at the resonating frequency.

2 Antenna Design

The slotted ground plane is made of a copper (annealed), above which the Beryllia
(99.5%) substrate having electrical permittivity of 6.5 has been placed. The substrate height is kept as 1.00 mm. A triangular shaped radiator made of Cu (annealed)
embedded with Sierpinski triangular slots has been placed. Figure 1 shows the complete view of the designed antenna. Parametric details have been provided in Table 1.

Parameters	Dimension (in mm)
L	36.41
W	27.52
L'	20.13
F_l	20.35
F_w	01.51
S_1	08.22
S_2	01.53
S_3	03.11
S_4	03.02

Table 1 Dimensions of proposed structure in mm

3 Genetic Algorithm

Those individuals who are capable of surviving in limited resources can exist in environment. For the survival of a certain species, it is necessary to adapt to the changing environment. Whether the species will survive or not is mainly governed by the genetic content. 'Gene' is the basic unit, which decides the probability of survival. 'Chromosomes,' which are nothing but a set of genes, play a major role in deciding the genetic characteristics of an individual. The 'survival of the fittest' theory states that the stronger species will survive and the weaker ones will be extinct. Genetic algorithm (GA) is motivated by the phenomenon of natural selection. Natural selection is a biological process of dominance of stronger individuals over weaker ones in a strong competitive environment. GA is analogous to the process of natural selection. It assumes that the solution to a problem is an individual, which can be represented by a parametric set. The genes are considered to be the parametric set that can be represented by a set in binary format. The 'fitness value' or the positive value reflects the level of 'goodness' and provides a potential solution to the problem [7, 8].

4 Application of GA in Antenna Design

In the process of genetic change, good quality offsprings are produced from a healthy chromosome, which provides a suitable solution to the problem. The objective value is closely related to the fitness value, which is particularly helpful in solving real-time problems that deal with optimization. A population of chromosomes is chosen for practical application purpose. Each and every problem has a unique population size. During the evolving process of genetic algorithm, the current population of chromosomes is used to create subsequent generations. These chromosomal set of the set of parents, usually regarded as the 'mating pool,' are chosen by specific procedures, for this process to be fruitful. These particular set of genes are selected

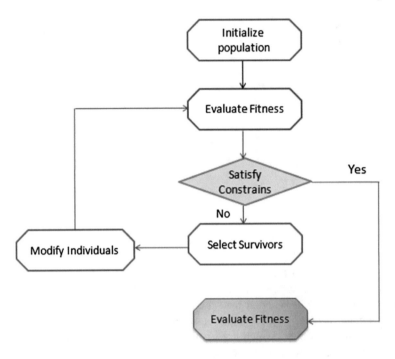

Fig. 2 Algorithmic phases of genetic algorithm

for production of offsprings in the subsequent generation. Following this process, it is therefore expected that mating of these superior quality genes will definitely yield good quality offsprings, having a higher probability of survival in the coming generations, following the process of 'survival of the fittest' [9, 10]. The evolution cycle is repeated until a desired point is reached. This objective can also be fulfilled by the total number of evolution cycles, i.e., number of computations, the standard of individual variation of different generations, or a particular fitness value, which can be graphically simplified with the help of a flowchart as displayed in Fig. 2.

This special property of genetic algorithm can be applied for the resonant frequency optimization of the proposed monopole antenna. The basic objective of this work is to optimize the return loss parameter, i.e., $|S_{11}|$ at the resonant frequency point. Using RCGA, the 'fitness function' can be calculated as,

Problem formulation:
Total generations: 300
Optimization parameter: S_{11}
Crossover probability: 0.5
Mutation probability: 0.075
Fitness function:

Fig. 3 Substrate effect on antenna return loss

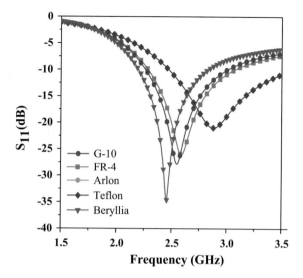

$$P(x) = \frac{1}{N} \sum_{i=1}^{N} Q(fi) \qquad (1)$$

$$Q(f_i) = \begin{cases} 15 & S11 < -15 \\ S11\,(fi) & S11 \geq -15 \end{cases} \qquad (2)$$

where $f_i = 2.45$ GHz.

The optimized antenna parameters have been provided in Table 1.

5 Results and Discussion

Figure 3 shows the frequency response of the antenna for some commonly available substrates, each characterized with different dielectric properties. The antenna has been found to resonate at 2.45 GHz with an impedance bandwidth of 650 MHz when Beryllia (95%) having an electrical permittivity of 6.5 is applied as the substrate material.

Figure 4 displays the plot of the optimized and unoptimized frequency response of the designed antenna. It has been shown that antenna return loss, i.e., $|S_{11}|$ has been effectively improved after genetic algorithmic optimization. For an unoptimized antenna the return loss is 20 dB, whereas for the optimized antenna the same is 35 dB.

Figure 5 shows the VSWR characteristics of the designed antenna structure. VSWR has been observed to be less than 2, which is no doubt, a good acceptable

Fig. 4 Optimized and
unoptimized return losses

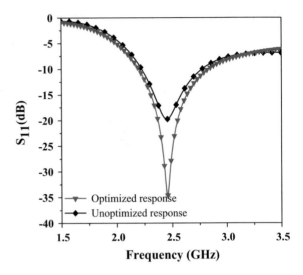

Fig. 5 VSWR
characteristics of proposed
monopole antenna

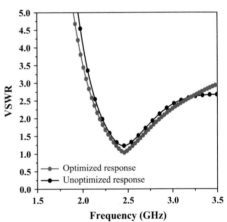

factor, better than non-optimized design. It has been observed that the VSWR has
been further reduced after optimization of the antenna parameters.

Gain versus frequency plot of the antenna has been shown in Fig. 6. It has been
increasing from 3.12 to 3.2 dBi linearly up to 2.5 GHz. At 2.45 GHz, the gain is
almost equal to 3.17 dBi. Gain is much higher in case of the optimized structure.

The antenna surface current distribution pattern and the radiation pattern are shown
in Fig. 7 and Fig. 8, respectively. Figure 7 shows that the surface current is concentrated near the Sierpinski triangular slots at the resonant frequency point, thereby
improving the return loss and realized antenna gain.

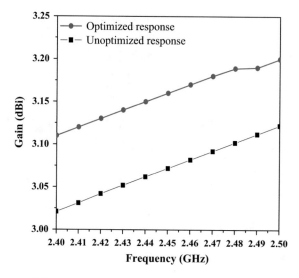

Fig. 6 Gain characteristics of proposed monopole antenna

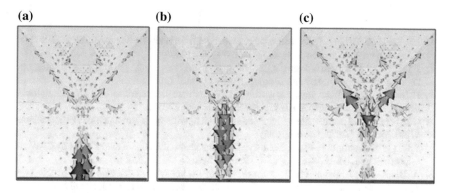

Fig. 7 Antenna surface current distribution pattern at: **a** 2.1 GHz, **b** 3.3 GHz, and **c** 2.45 GHz

6 Conclusion

In this article, a compact fractal monopole antenna with RCGA optimized return loss has been investigated. The peak resonant frequency of the antenna is exactly at 2.45 GHz with an impedance bandwidth of 650 MHz. The antenna finds its application in 2.4 GHz (2.41–2.48 GHz) WLAN band with a maximum realized gain of 3.36 dBi at the resonating frequency. The frequency response as well as the radiation characteristics of the proposed monopole antenna is therefore better than conventional patches, which has been verified and justified from the above investigation.

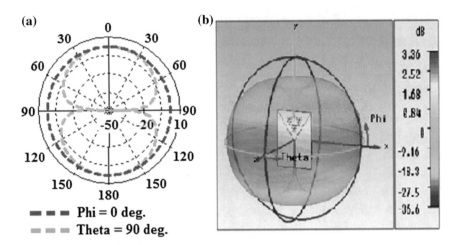

Fig. 8 Antenna radiation pattern at 2.45 GHz **a** 2-D view and **b** 3-D view

References

1. J.Y. Jan, C.Y. Hsiang, Wideband coplanar waveguide-fed slot antenna for DCS, PCS, 3 G and Bluetooth bands. Electron. Lett. **42**(24), 1377–1378 (2006)
2. J.Y. Chiou, J.Y. Sze, K.L. Wong, A broadband coplaner waveguide-fed strip loaded square slot antenna. IEEE Trans. Antennas Propag. **51**(4), 719–721 (2003)
3. H.D. Chen, Broadband coplanar wave guide-Fed square slot antennas with a widened tuning stub. IEEE Trans. Antennas Propag. **51**(4), 1982–1986 (2003)
4. C.J. Wang, J.J. Lee, A pattern-frequency-dependent wide-band slot antenna. IEEE Antennas Wirel. Propag. Lett. **5**, 65–68 (2006)
5. S.W. Qu, C. Ruan, B.Z. Wang, Bandwidth enhancement of wide-slot antenna fed by CPW and microstrip line. IEEE Antennas Wirel. Propag. Lett. **5**, 15–17 (2006)
6. A. Bhattacharya, An analytical approach to study the behavior of defected patch structures, in *Emerging Trends in Computing and Communication*. Lecture Notes in Electrical Engineering, vol. 238 (Springer, 2014), pp. 431–433
7. L. Davis, *Genetic Algorithms and Simulated Annealing* (Pitman, London, U.K., 1987)
8. K.A. De Jong, An analysis of the behavior of a class of genetic adaptive systems, Doctoral Dissertation, Univ. of Michigan, Ann Arbor, Mich. (1975)
9. K.A.D. Jong, Genetic algorithms: a 10 year perspective, in Proceedings of the International Conference on Genetic Algorithms Application, Hillsdale, NJ (1985)
10. L. Bukatova, Y.V. Gulyaev, From genetic algorithms to evolutionary computer, in *Proceedings of the 5th International Conference Genetic Algorithms, Urbana, IL, July 1993* (1993), pp. 614–617

Improvement of Radiation Performances of Butler Matrix-Fed Antenna Array System Using 4 × 1 Planar Circular EBG Units

Wriddhi Bhowmik, Surajit Mukherjee, Vibha Rani Gupta, Shweta Srivastava and Laxman Prasad

Abstract Excitation of surface waves in the microstrip patch antenna at higher frequencies degrades the overall radiation performances. To overcome this problem, electromagnetic band gap (EBG) structures can be used. In this paper, the improvement of radiation performances of a Butler matrix-fed antenna array system has been obtained by incorporating 4 × 1 planar circular EBG units in the form of metallic patches between the array elements as well as beneath the radiating patches as circular slots. Finally, symmetric gain has been obtained for all the radiated beams at 14.5 GHz.

Keywords Butler matrix array · Surface waves · Electromagnetic band gap (EBG) structure

1 Introduction

Minimization of interference and multipath fading makes improve the performance of high-speed wireless communication. Multiple beam forming antenna systems (M-BFAS) can be used for this purpose. It generates very directive and narrow beams in different angular directions. These directive beams reduce the possibility of interference as well as fading. The Butler matrix array is the suitable option of beam forming network (BFN) as very few components are required for matrix

W. Bhowmik (✉) · S. Mukherjee
Haldia Institute of Technology, Haldia, West Bengal, India
e-mail: bhowmikwriddhi@gmail.com

V. Rani Gupta
Birla Institute of Technology, Mesa, Ranchi, Jharkhand, India

S. Srivastava
Jaypee Institute of Information Technology, Noida, Uttar Pradesh, India

L. Prasad
Raj Kumar Goel Institute of Technology, Ghaziabad, India

© Springer Nature Singapore Pte Ltd. 2017
J. Bhaumik et al. (eds.), *Communication, Devices, and Computing*, Lecture Notes in Electrical Engineering 470, https://doi.org/10.1007/978-981-10-8585-7_10

implementation [1]. Also, the matrix introduces low loss to the microwave signal transmission.

The Butler matrix array comprises of basic microwave devices, such as 3 and 0 dB coupler, 45° phase shifter [2]. The implementation of Butler matrix array has been made by microstrip technique to make the system compact as well as inexpensive. Though most of the times researchers used rectangular patch microstrip antenna as radiating element in M-BFAS to generate directive beams in different angular directions, the surface wave excitation at higher frequencies in the patch antenna degrades its radiation efficiency as well as the radiation performances of the whole antenna array system.

EBG structures can be used to improve the radiation performances as it restricts the surface wave excitation. Previously, some complex structures of Butler matrix-fed antenna array system have been developed by using EBG to minimize the mutual coupling as well as enhance the gain of radiated beams [3–5]. This paper introduces a simple structure of antenna array system, incorporated with 4 × 1 planar circular EBG units to improve the radiation performances of the system.

FR-4 epoxy with relative permittivity (ε_r) of 4.4, dissipation factor (tan δ) of 0.025, and height (h) of 1.6 mm has been used for design implementation. This material is inexpensive and also easily available. These features of FR-4 epoxy make it more attractive for frequent use though it suffers from high dielectric loss at higher frequencies as well as surface wave excitation. Again the thickness (h) of the substrate is another reason of surface wave excitation in the same. Hence, the use of this material at Ku band applications is not a very suitable option. Besides all these disadvantages of this material, it has also been observed that to reduce cost of the system, some researchers have been used the FR-4 epoxy as design material in Ku band applications [6–8]. In this paper, to make the proposed antenna array system compact and inexpensive, microstrip technique and FR-4 epoxy have been used.

1.1 Butler Matrix Array

Generally, a Butler matrix array is a $N \times N$ matrix ($N = 2^n$). The output ports of Butler matrix array feed N element array, which produces N directive beams in different angular directions [9]. Distinct phase differences are obtained between the output ports of the matrix, while input ports get excited by a signal. The direction of the radiated beams depends upon these phase differences. The beam direction θ can be calculated as follows [4]

$$\theta = \cos^{-1}\left[\frac{\pm(2n-1)\lambda_0}{8d}\right] \tag{1}$$

In Eq. (1), $n = 1, 2, ..., \lambda_0$ is wavelength and d is the center to center distance of antenna elements.

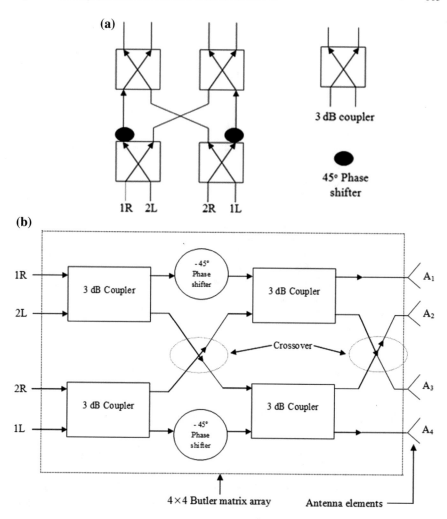

Fig. 1 **a** Schematic of 4 × 4 Butler matrix array and **b** M-BFAS

Figure 1a and b, respectively, represents the schematic of 4 × 4 Butler matrix array and multiple beam forming antenna system (M-BFAS) [10].

The power will flow through the Butler matrix network, while any input port of the matrix gets excited, and finally, one-fourth of the input power is fed to the antenna elements. The cumulative effect of four-element array (shown in Fig. 1b) will produce a directive beam. The excitation of input port named 1R (first right) of M-BFAS (shown in Fig. 1b) yields the directive beam 1R, which should be in the angular range of 0° to 30°, as shown in Fig. 2. Similarly, the 1L, 2R, and 2L beams should be in the angular range of −30° to 0°, 30° to 60°, and −60° to −30°, respectively, while the ports 1L, 2R, and 2L of M-BFAS get excited.

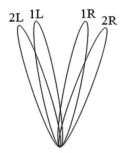

Fig. 2 Proposed radiation pattern of M-BFAS

(a) **(b)**

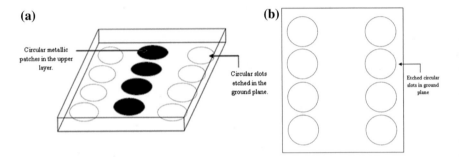

Fig. 3 **a** Complete structure of proposed EBG unit and **b** ground plane

1.2 Electromagnetic Band Gap Structure

Minimization of surface wave excitation can be achieved by EBG structures. The integration of EBG structures with microstrip patch antenna is simple and inexpensive. Periodic structure of air holes through substrate, periodic metal patches with grounding vias, uniplanar compact EBG (UC-EBG) structure, spiral EBG structure, periodic planar rectangular metal patches or circular slots etched in ground plane are different forms of EBG structure [11]. In this paper, simple configuration of EBG has been used for gain enhancement.

2 Design of 4 × 1 Planar Circular EBG Unit

A layout of proposed 4 × 1 planar circular EBG unit is presented in Fig. 3.

It is evident from Fig. 4 that S_{11} of −14.35 dB has been achieved at 14.5 GHz. A stop band of 4.3 GHz has been obtained by the proposed structure as shown in Fig. 4. So it can be assumed that after implementation of the proposed 4 × 1 EBG unit in the antenna array system, better radiation performance might be obtained.

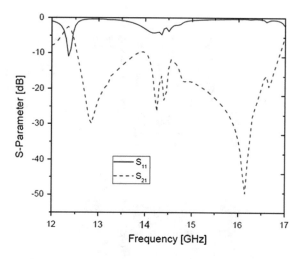

Fig. 4 S-parameter characteristics

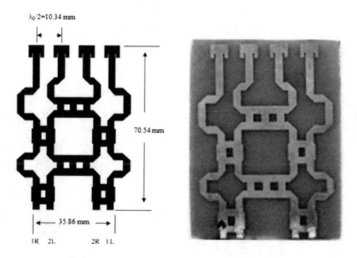

Fig. 5 Layout and fabricated structure at 14.5 GHz

3 Design of Butler Matrix-Fed Antenna Array System Incorporated with 4 × 1 EBG Units

The conventional structure is presented in Fig. 5.

The conventional structure of antenna array system has been modified by incorporating the 4 × 1 EBG units as metallic patches between the array elements as well as beneath the radiating patches as circular slots. The layout and fabricated structure are shown in Fig. 6 and Fig. 7, respectively.

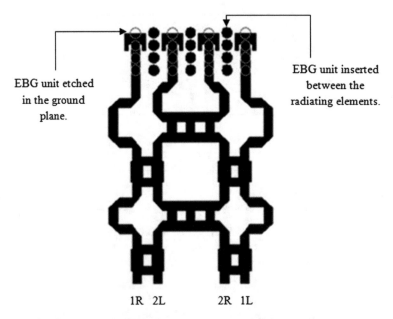

EBG unit etched
in the ground
plane.

EBG unit inserted
between the
radiating elements.

1R 2L   2R 1L

Fig. 6 Layout of proposed M-BFAS

(a)   **(b)**

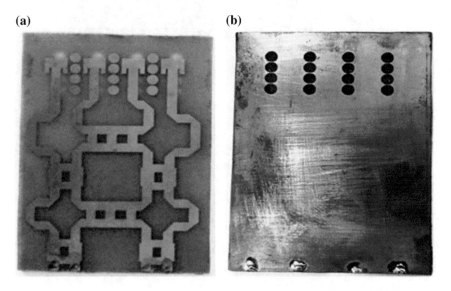

Fig. 7 Fabricated structure: **a** top metallization layer and **b** ground plane

The simulated and measured S_{11} are well below of -15 dB at 14.5 GHz as shown in Fig. 8. The average isolation level (S_{31}, S_{41}, and S_{42}) of -29 dB and (S_{21}, S_{32}, and S_{43}) -17.7 dB has been achieved as observed in Fig. 9; hence, the interference between

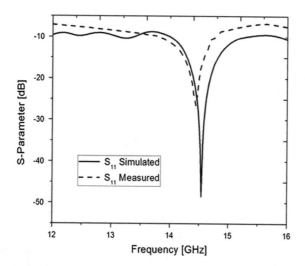

Fig. 8 Simulated and measured S_{11} of proposed M-BFAS

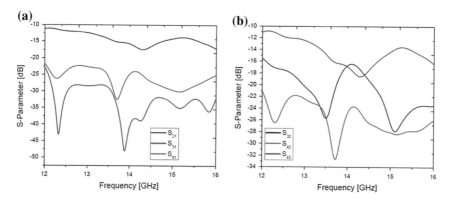

Fig. 9 Simulated isolation: **a** S_{21}, S_{31}, S_{41} and **b** S_{32}, S_{42}, S_{43}

the ports are low and in satisfactory level. Figure 10 represents the comparative analysis of normalized E-plane radiation performances of conventional and modified M-BFAS. A comparative study of first null point levels of both the M-BFAS is listed in Table 1.

First null point levels of every radiated beams of proposed M-BFAS have been improved compared to the conventional design as observed in Table 1. Major beam direction of proposed M-BFAS is in good agreement with conventional system as observed in Table 2. A comparative study of radiated major beam gains of both the conventional and proposed systems is reported in Table 3. The gain of 1R beam is similar for both the design. An increment of 2.7 dB, 1.73 dB, and 0.51 dB has been obtained in simulation for the 2L, 2R, and 1L beams, respectively. Also, in

Fig. 10 Comparative study
of normalized E-plane
patterns of the conventional
and proposed M-BFAS at
14.5 GHz for different input
ports excitation: **a** 1R, **b** 2L,
c 2R, and **d** 1L

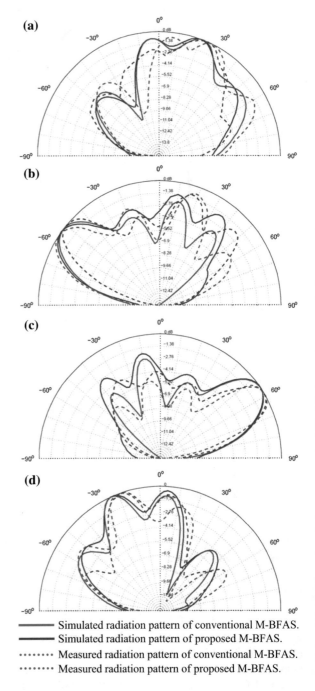

Simulated radiation pattern of conventional M-BFAS.
Simulated radiation pattern of proposed M-BFAS.
Measured radiation pattern of conventional M-BFAS.
Measured radiation pattern of proposed M-BFAS.

Table 1 Comparative study of first null point levels

Beams	Conventional design (Fig. 5)	Proposed design (Fig. 7)
	First null points (dB)	First null points (dB)
1R	−1.7 and −2.77	−2.11 and −3.8
2L	−3.82	−4.23
2R	−5	−5.57
1L	−3.17 and −2.42	−4.62 and −3.2

Table 2 Comparative study of beam directions

Beams	Conventional design (Fig. 5)	Proposed design (Fig. 7)
	Measured beam directions	Measured beam directions
1R	+22°	+26°
2L	−50°	−56°
2R	+58°	+60°
1L	−18°	−20°

Table 3 Comparative study of radiated major beam gain

Beams	Conventional design (Fig. 5)		Proposed design (Fig. 7)	
	Simulated gain (dB)	Measured gain (dB)	Simulated gain (dB)	Measured gain (dB)
1R	4.50	3.77	4.52	4.24
2L	1.45	1.23	4.15	3.81
2R	2.62	2.22	4.35	4.01
1L	4.32	4.23	4.83	4.78

Table 4 Comparative study of half-power beam width (HPBW)

Beams	HPBW	
	Conventional design (Fig. 5)	Modified design (Fig. 7)
1R	53°	47°
2L	32°	33°
2R	31°	31°
1L	30°	25°

measurement of 0.47 dB, 2.58 dB, 1.79 dB, and 0.55 dB, gain enhancement has been achieved for 1R, 2L, 2R, and 1L beams, respectively. It is observed from Fig. 10b, c that more power has been transferred to the side lobes of 2L and 2R beams of conventional design, whereas less power transferred to the side lobes of 2L and 2R beams of modified design. Hence, the gain improvement of 2.7 dB and 1.73 dB has been obtained for 2L and 2R beams of modified design, respectively, as compared to the conventional system. The comparative study of half-power beam width of radiated major beams of the conventional and proposed M-BFAS is reported in Table 4.

The integration of planar circular EBG unit in the conventional antenna array system makes the radiation performance better. It has been observed that in modified design, the first null points of the major beams are improved than the conventional design. The comparative study of HPBW of different major beams shows that some improvements have been achieved in HPBW of the radiated beams in the experiment. This integration of 4 × 1 EBG unit with the conventional M-BFAS results in the improvement of HPBW of 6° and 5° for the beams 1R and 1L, respectively, as observed in Table 4. The HPBW of 2L and 2R beam is similar to the conventional design. The gain of radiated beams of the conventional system is asymmetric; comparatively, lower gain has been obtained for 2L and 2R beams than the 1R and 1L beams as observed in Table 3. Overall experiments show that by using the 4 × 1 planar circular EBG unit in the conventional design, the symmetric gain for different major beams has been obtained.

The electromagnetic band gap structure generates the stop band region wherein the surface wave excitation is restricted. Hence, more input power will be transmitted to the antenna array at the operating frequency and increases the gain of major beams while implementing the EBG unit. The wider stop band region, approximately 4.3 GHz, has been obtained by the 4 × 1 EBG unit. This might be the reason of getting gain enhancement as well as the symmetricity in gain of radiated beams of the proposed M-BFAS while using the 4 × 1 EBG unit.

4 Conclusion

To improve the radiation performances of conventional Butler matrix-fed antenna array system, a planar circular 4 × 1 EBG units have been integrated with the system. The HPBW and first null points of radiated beams have been improved as well as symmetric gain for all the major beams has been obtained by the proposed M-BFAS.

References

1. W. White, Pattern limitations in multiple-beam antennas. IRE trans. Antennas Propag. **10**, 430–436 (1962)
2. C.-H. Tseng, C.-J. Chen, T.-H. Chu, A low-cost 60-GHz switched-beam patch antenna array with Butler matrix network. IEEE Antennas Wirel. Propag. Lett. **07**, 432–435 (2008)
3. E.R. Iglesias, O.Q. Teruel, L.I. Sanchez, Mutual coupling reduction in patch antenna array by using a planar EBG structure and a multilayer dielectric substrate. IEEE Trans. Antennas Propag. **56**, 1648–1655 (2008)
4. G.E. Dominguez, J.-M. Gonzalez, P. Padilla, M.S. Castaner, Mutual coupling reduction using EBG in steering antennas. IEEE Antennas Wirel. Propag. Lett. **11**, 1265–1268 (2012)
5. M.M. Aldemerdash, A.A. Mitkees, H.A. Elmikati, Effect of mutual coupling on the performance of four elements microstrip antenna array fed by a Butler matrix, in *29th National Radio Science Conference, Egypt* (2012), pp. 127–140

6. S. Chakraboty, S. Srivastava, Ku band annular ring antenna on different PBG structure. Int. J. Mod. Eng. Res. **02**, 4726–4731 (2012)
7. A.A. Jamali, A. Gaafar, A.A. Abd Elaziz, Finite different ground shapes printed spiral antennas for multi wide band applications using PPPC feeding scheme, in *Progress in Electromagnetic Research Proceedings, China* (2011), pp. 224–229
8. W. Bhowmik, S. Srivastava, L. Prasad, Design of a low cost 4 × 4 Butler matrix fed antenna array partially loaded with substrate integrated wavegudie. Int. J. Microwave Opt. Technol. **09**, 227–236 (2014)
9. S.R. Ahmad, F.C. Seman, 4-port Butler matrix for switched multibeam antenna array, in *Asia Pacific Conference on Applied Electromagnetics, Malaysiya* (2005), pp. 69–73
10. W. Bhowmik, V.R. Gupta, S. Srivastava, L. Prasad, Gain enhancement of Butler matrix fed antenna array system by using planar circular EBG units, in *IEEE International Conference on Signal Processing and Communication, India* (2015), pp. 183–188
11. R. Garg, P. Bharatia, I. Bahl, A. Ittipiboon, *Microstrip Antenna Design Handbook* (Artech House, Norwood, MA, 2001)

Improving Security of SPN-Type Block Cipher Against Fault Attack

Gitika Maity, Sunanda Jana, Moumita Mantri and Jaydeb Bhaumik

Abstract Differential fault attack (DFA) is the most popular technique often used to attack physical implementation of block cipher by introducing a computational error. In this paper, a new modified SPN-type architecture has been proposed which provides better resistance against fault attack compared to AES. The proposed architecture is similar to AES architecture except round key mixing function. A nonlinear vectorial Boolean function called Nmix is used to mix the round key with round output, which is a 16-bit mixing operation. 128-bit 10th round key is retrieved using 24 faulty–fault-free ciphertext pairs by injecting a fault at the input of 9th round, before sub-byte operation. It needs computation complexity of 2^{53} which is much greater compared to original AES to find 128 bit of 10th round key.

Keywords Fault attack · Block cipher · Substitution and permutation network · Nonlinear Boolean function

1 Introduction

Nowadays, several side-channel attacks are popularly used to attack cryptographic algorithms, like fault attack, power consumption analysis attack, electromagnetic radiation attack, timing attack, cache attack. These attacks can be easily mounted by the intruder by using any low-cost instrument. These side-channel attacks amplify and evaluate leaked information with the help of statistical methods and are often much more powerful than the classical cryptanalysis. Among these attacks, differential fault attack (DFA) is one of the most popular side-channel attacks in the field of

G. Maity (✉) · S. Jana
Department of CSE, Haldia Institute of Technology, Haldia, India
e-mail: Gitika.Maity@gmail.com

M. Mantri
Department of IT, Haldia Institute of Technology, Haldia, India

J. Bhaumik
Department of ECE, Haldia Institute of Technology,
Haldia 721657, West Bengal, India

© Springer Nature Singapore Pte Ltd. 2017
J. Bhaumik et al. (eds.), *Communication, Devices, and Computing*, Lecture Notes in Electrical Engineering 470, https://doi.org/10.1007/978-981-10-8585-7_11

115

cryptography. Fault attack on crypto-hardware was first introduced by Boneh et al. [1, 2] from Bellcore. Later on the concept of Differential fault attack (DFA) was introduced by Biham and Shamir [3] on the Data Encryption Standard (DES).

The US National Institute of Standards and Technology (NIST) selected Rijndael [4] as the Advanced Encryption Standard (AES) in 2000, and it has been adopted as a worldwide standard for symmetric key encryption. Till date, fault-based attack on advanced encryption algorithm has lowest computational complexity compared to all other attacks. Also presently very low-cost methods are used for fault injection such as variation of supply voltages, clock glitches, temperature variation, UV light radiations. Finding key from algebraic equation was accepted in [5]. Inducing faults at byte level to the input of 9th round of AES for DFA was reported in [6], where 250 faulty ciphertexts are needed to recover the key. Differential fault attack against AES was analyzed by Blomer and Seifert [7], and they showed that by injecting fault at byte level in the 8th round and 9th round the attacker can derive the key using 40 ciphertexts. In [8], it is given that by introducing fault at byte level in the 8th round and 9th round of AES-128 algorithm, attacker can easily recover the total key using two faulty ciphertexts. A fault-based attack on MDS-AES has been reported in [9] where from one faulty encryption actual key can be retrieved with a brute-force search of complexity 2^{16}. A survey on fault attack against AES and their countermeasures have been discussed in [10].

In this paper, a fault-based side-channel attack on a SPN-type block cipher has been analyzed where XOR operation in add round key step is replaced by a nonlinear Nmix−16-bit function [11]. In this proposed attack, a random byte fault has been injected at the first byte of the 9th round. At the end of the experiment, it is observed that 48-bit key can be recovered by analyzing 6 faulty ciphertexts. The attacker has to search for 2^{48} which does not require any brute-force to find the final key. To find 128-bit keys, it needs 24 faulty and fault-free ciphertext pairs. Here, the attack has a computation complexity of 2^{53} when fault is injected before 9th round.

The rest of the paper is organized as follows: Section 2 gives a description of the proposed SPN-type block cipher algorithm. Fault attack on ninth round of the proposed SPN-Type architecture has been described in Sect. 3 and finally, the paper is concluded in Sect. 4.

2   Description of SPN-Type Block Cipher Algorithm

The AES-Rijndael encryption algorithm is composed of rounds. There are 10 rounds for AES-128. For first 9 rounds, each round consists of four steps called SubByte, ShiftRows, MixColumn, and AddRoundKey. In the last round, the step MixColumn is omitted. The proposed SPN-type architecture is a modified version of AES-Rijndael algorithm. The description of AES-Rijndael algorithm is available in [4]. In this section, a brief description of the proposed SPN-type block cipher algorithm is provided. In this algorithm, key size is 128 bits and block size is also 128 bits. The 128-bit message block is arranged as 4×4 array of bytes which is called as state matrix. The elements of the matrix are represented by variables s_{ij} where $0 \leq i \leq 3$ and $0 \leq j \leq 3$, where i, j denoting the row and column indexes of the state matrix. Number of round is 10 for 128-bit key, and in each round, keys are generated from the initial key. Round keys are generated from key scheduling algorithm. At last after round 10, the ciphertext is generated. The SPN-type block cipher algorithm consists of the following steps: SubByte, ShiftRows, MixColumn, and Round Key Mixing(AddKeyNmix16bit) similar to AES except Round Key Mixing. A nonlinear function AddKeyNmix16bit is used in AddRoundKey step for encryption. For decryption, the inverse Nmix(AddKeyINVNmix16bit) is used. Details about Nmix and INmix may be found in [11]. All the steps involved in encryption and decryption are shown in Fig. 1.

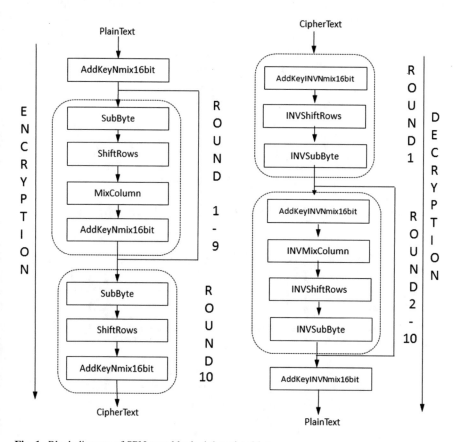

Fig. 1 Block diagram of SPN-type block cipher algorithm

3   Fault Attack on Ninth Round of the Proposed SPN-Type Architecture

In this section, a single random byte nonzero fault is introduced at the first byte of the input of 9th round SubByte step. The propagation of fault is shown in Fig. 2. The fault is g_1 at 9th round SubByte step. After SubByte step, it changes to g_1'. After operation of ShiftRows, the fault g_1' is confined to same position. In MixColumn, the fault is distributed into 4 bytes of 1st column. After ShiftRows, the faults are distributed into the four bytes. After the AddKeyNmix16bit operation, the fault is distributed into 6 bytes. The mixing operation we perform here is of 16 bits. The fault is considered to be in lower 8 bits and affect the higher 8 bits as well. Assuming $CT1$ is a fault-free ciphertext and $CT2$ is the corresponding faulty ciphertext, then the following ciphertexts are

$$CT1 = \begin{pmatrix} x_0 & x_4 & x_8 & x_{12} \\ x_1 & x_5 & x_9 & x_{13} \\ x_2 & x_6 & x_{10} & x_{14} \\ x_3 & x_7 & x_{11} & x_{15} \end{pmatrix}$$

$$CT2 = \begin{pmatrix} (x_0 + D11) & x_4 & x_8 & x_{12} \\ (x_1 + D_{12}) & x_5 & x_9 & (x_{13} + D_2) \\ x_2 & x_6 & (x_{10} + D_{31}) & x_{14} \\ x_3 & (x_7 + D_4) & (x_{11} + D_{32}) & x_{15} \end{pmatrix}$$

We consider the associated key matrices K_9 and K_{10} in 9th and 10th rounds as follows

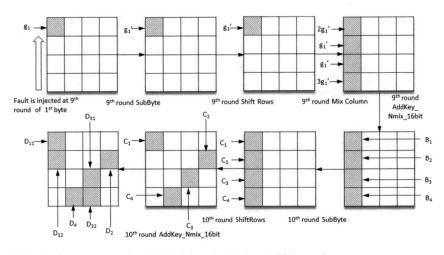

Fig. 2 Fault propagation when fault is injected at the input of 9th round

$$K_9 = \begin{pmatrix} k_{90} & k_{94} & k_{98} & k_{912} \\ k_{91} & k_{95} & k_{99} & k_{913} \\ k_{92} & k_{96} & k_{910} & k_{914} \\ k_{93} & k_{97} & k_{911} & k_{915} \end{pmatrix}$$

$$K_{10} = \begin{pmatrix} k_{100} & k_{104} & k_{108} & k_{1012} \\ k_{101} & k_{105} & k_{109} & k_{1013} \\ k_{102} & k_{106} & k_{1010} & k_{1014} \\ k_{103} & k_{107} & k_{1011} & k_{1015} \end{pmatrix}$$

In the following equations, INVmixLSB is doing 16-bit Nmix operation and it gives only the lower byte. Similarly, INVmixMSB gives only the higher byte. ISB is the inverse sub-byte operation. From the fault pattern, following equations can be constructed

$$(INVmixLSB(ISB(INVmixLSB(x_0, x_1, k_{100}, k_{101})),$$
$$ISB(INVmixMSB(x_0, x_1, k_{100}, k_{101})), k_{90}, k_{91})) \oplus$$
$$(INVmixLSB(ISB(INVmixLSB(x_0 + D_{11}, x_1 + D_{12}, k_{100}, k_{101}))),$$
$$ISB(INVmixLSB(x_0 + D_{11}, x_1 + D_{12}, k_{100}, k_{101}))), k_{90}, k_{91}) =$$
$$2[(INVmix(ISB(INVmix(x_{13}, k_{1013})), k_{91})) \oplus$$
$$(INVmix(ISB(INVmix(x_{13} + D_2), k_{1013})), k_{91})] \tag{1}$$

$$(INVmixLSB(ISB(INVmixLSB(x_{10}, x_{11}, k_{1010}, k_{1011})),$$
$$ISB(INVmixMSB(x_{10}, x_{11}, k_{1010}, k_{1011})), k_{91}, k_{92})) \oplus$$
$$(INVmixLSB(ISB(INVmixLSB(x_{10} + D_{31}, x_{11} + D_{32}, k_{1010}, k_{1011}))),$$
$$ISB((INVmixMSB(x_{10} + D_{31}, x_{11} + D_{32}, k_{1010}, k_{1011}))), k_{91}, k_{92}) =$$
$$(INVmix(ISB(INmix(x_{13}, k_{1013})), k_{91})) \oplus$$
$$(INVmix(ISB(INVmix(x_{13} + D_2, k_{1013}))), k_{91}) \tag{2}$$

$$(INVmix(ISB(INmix(x_7, k_{107})), k_{93})) \oplus$$
$$(INVmix(ISB(INVmix(x_7 + D_4, k_{107}))), k_{93}) =$$
$$3[(INVmixLSB(ISB(INVmixLSB(x_{10}, x_{11}, k_{1010}, k_{1011})),$$
$$ISB((INVmixMSB(x_{10}, x_{11}, k_{1010}, k_{1011})), k_{92}, k_{93})) \oplus$$
$$(INVmixLSB(ISB(INVmixLSB(x_{10} + D_{31}), x_{11} + D_{32}, k_{1010}, k_{1011})),$$
$$ISB((INVmixMSB(x_{10} + D_{31}, x_{11} + D_{32}, k_{1010}, k_{1011})), k_{92}, k_{93})) \tag{3}$$

In Eqs. (1), (2), and (3), the keys k_{100}, k_{101}, k_{107}, k_{1010}, k_{1011}, and k_{1013} are the 10th round keys and k_{90}, k_{91}, k_{92}, and k_{93} are the 9th round keys. Only 48-bit keys are recovered.

Similarly if the attacker injects a nonzero random fault at 5th byte, then another 48-bit keys are obtained. Assuming $CT3$ is a fault-free ciphertext and $CT4$ is the corresponding faulty ciphertext, then the following ciphertexts are

$$CT3 = \begin{pmatrix} x_0 & x_4 & x_8 & x_{12} \\ x_1 & x_5 & x_9 & x_{13} \\ x_2 & x_6 & x_{10} & x_{14} \\ x_3 & x_7 & x_{11} & x_{15} \end{pmatrix}$$

$$CT4 = \begin{pmatrix} x_0 & (x_4 + D_{51}) & x_8 & x_{12} \\ (x_1 + D_6) & (x_5 + D_{52}) & x_9 & x_{13} \\ x_2 & x_6 & x_{10} & (x_{14} + D_{71}) \\ x_3 & x_7 & (x_{11} + D_8) & (x_{15} + D_{72}) \end{pmatrix}$$

From the fault propagation, following equations are constructed

$$\begin{aligned}
(INVmixLSB(&ISB(INVmixLSB(x_4, x_5, k_{104}, k_{105})), \\
&ISB((INVmixMSB(x_4, x_5, k_{104}, k_{105})), k_{94}, k_{95})) \oplus \\
(INVmixLSB(&ISB(INVmixLSB(x_4 + D_{51}, x_5 + D_{52}, k_{104}, k_{105}))), \\
&ISB((INVmixMSB(x_4 + D_{51}, x_5 + D_{52}, k_{104}, k_{105}))), k_{94}, k_{95}) = \\
&3[(INVmix(ISB(INVmix(x_{11}, k_{1011})), k_{97})) \oplus \\
&(INVmix(ISB(INVmix(x_{11} + D_8, k_{1011}))), k_{97}))]
\end{aligned} \quad (4)$$

$$\begin{aligned}
(INVmix(ISB(&INVmix(x_1, k_{101})), k_{95})) \oplus \\
(INVmix(ISB(&INVmix(x_1 + D_6, k_{101}))), k_{95})) = \\
&2[(INVmixLSB(ISB(INVmixLSB(x_{14}, x_{15}, k_{1014}, k_{1015})), \\
&ISB((INVmixMSB(x_{14}, x_{15}, k_{1014}, k_{1015})), k_{96}, k_{97})) \oplus \\
&(INVmixLSB(ISB(INVmixLSB(x_{14} + D_{71}, x_{15} + D_{72}, k_{1014}, k_{1015}))), \\
&ISB((INVmixMSB(x_{14} + D_{71}, x_{15} + D_{72}, k_{1014}, k_{1015}))), k_{96}, k_{97}]
\end{aligned} \quad (5)$$

$$\begin{aligned}
(INVmixLSB(&ISB(INVmixLSB(x_{14}, x_{15}, k_{1014}, k_{1015})), \\
&ISB((INVmixMSB(x_{14}, x_{15}, k_{1014}, k_{1015})), k_{96}, k_{97})) \oplus \\
(INVmixLSB(&ISB(INVmixLSB(x_{14} + D_{71}, x_{15} + D_{72}, k_{1014}, k_{1015}))), \\
&ISB((INVmixMSB(x_{14} + D_{71}, x_{15} + D_{72}, k_{1014}, k_{1015}))), k_{96}, k_{97}) = \\
&(INVmix(ISB(INVmix(x_{11}, k_{1011})), k_{97})) \oplus \\
&(INVmix(ISB(INVmix(x_{11} + D_8, k_{1011}))), k_{97}))
\end{aligned} \quad (6)$$

From Eqs. (4), (5), and (6), only 48-bit 10th round keys k_{101}, k_{104}, k_{105}, k_{1011}, k_{1014}, and k_{1015} are obtained and k_{94}, k_{95}, k_{96}, and k_{97} 9th round keys are recovered. If fault is injected at 9th byte of 9th round SubByte step, then another 48-bit keys are recovered. Assuming $CT5$ is a fault-free ciphertext and $CT6$ is the corresponding faulty ciphertext, then the following ciphertexts are

$$CT5 = \begin{pmatrix} x_0 & x_4 & x_8 & x_{12} \\ x_1 & x_5 & x_9 & x_{13} \\ x_2 & x_6 & x_{10} & x_{14} \\ x_3 & x_7 & x_{11} & x_{15} \end{pmatrix}.$$

$$CT6 = \begin{pmatrix} x_0 & x_4 & (x_8 + D_{91}) & x_{12} \\ x_1 & (x_5 + D_{10}) & (x_9 + D_{92}) & x_{13} \\ (x_2 + D_{111}) & x_6 & x_{10} & x_{14} \\ (x_3 + D_{112}) & x_7 & x_{11} & (x_{15} + D_{12}) \end{pmatrix}$$

$$
\begin{aligned}
&(INVmix(ISB(INVmix(x_5, k_{105})), k_{99})) \oplus \\
&(INVmix(ISB(INVmix(x_5 + D_{10}, k_{105}))), k_{99})) = \\
&3[(INVmixLSB(ISB(INVmixLSB(x_8, x_9, k_{108}, k_{109})), \\
&ISB((INVmixMSB(x_8, x_9, k_{108}, k_{109})), k_{98}, k_{99})) \oplus \\
&(INVmixLSB(ISB(INVmixLSB(x_8 + D_{91}, x_9 + D_{92}, k_{108}, k_{109}))), \\
&ISB((INVmixMSB(x_8 + D_{91}, x_9 + D_{92}, k_{108}, k_{109}))), k_{98}, k_{99})] \quad (7)
\end{aligned}
$$

$$
\begin{aligned}
&(INVmix(ISB(INVmix(x_5, k_{105})), k_{99})) \oplus \\
&(INVmix(ISB(INVmix(x_5 + D_{10}, k_{105}))), k_{99})) = \\
&2[(INVmixLSB(ISB(INVmixLSB(x_2, x_3, k_{102}, k_{103})), \\
&ISB((INVmixMSB(x_2, x_3, k_{102}, k_{103})), k_{910}, k_{911})) \oplus \\
&(INVmixLSB(ISB(INVmixLSB(x_2 + D_{111}, x_3 + D_{112}, k_{102}, k_{103}))), \\
&ISB((INVmixMSB(x_2 + D_{111}, x_3 + D_{112}, k_{102}, k_{103}))), k_{910}, k_{911})] \quad (8)
\end{aligned}
$$

$$
\begin{aligned}
&(INVmixLSB(ISB(INVmixLSB(x_8, x_9, k_{108}, k_{109})), \\
&ISB((INVmixMSB(x_8, x_9, k_{108}, k_{109})), k_{98}, k_{99})) \oplus \\
&(INVmixLSB(ISB(INVmixLSB(x_8 + D_{91}, x_9 + D_{92}, k_{108}, k_{109}))), \\
&ISB((INVmixMSB(x_8 + D_{91}, x_9 + D_{92}, k_{108}, k_{109}))), k_{98}, k_{99}) = \\
&(INVmix(ISB(INVmix(x_{15}, k_{1015})), k_{911})) \oplus \\
&(INVmix(ISB(INVmix(x_{15} + D_{12}, k_{1015}))), k_{911})) \quad (9)
\end{aligned}
$$

From Eqs. (7), (8), and (9), only 48-bit 10th round keys $k_{102}, k_{103}, k_{105}, k_{108}, k_{109}$, and k_{1015} are obtained and k_{98}, k_{99}, k_{910}, and k_{911} 9th round keys are recovered. Similarly if the attacker injects a nonzero random fault at 13th byte, then another 48-bit keys are obtained. Assuming $CT7$ is a fault-free ciphertext and $CT8$ is the corresponding faulty ciphertext, then the following ciphertexts are

$$
CT7 = \begin{pmatrix} x_0 & x_4 & x_8 & x_{12} \\ x_1 & x_5 & x_9 & x_{13} \\ x_2 & x_6 & x_{10} & x_{14} \\ x_3 & x_7 & x_{11} & x_{15} \end{pmatrix}
$$

$$
CT8 = \begin{pmatrix} x_0 & x_4 & x_8 & (x_{12} + D_{131}) \\ x_1 & x_5 & (x_9 + D_{14}) & (x_{13} + D_{132}) \\ x_2 & (x_6 + D_{151}) & x_{10} & x_{14} \\ (x_3 + D_{16}) & (x_7 + D_{152}) & x_{11} & x_{15} \end{pmatrix}
$$

$$
\begin{aligned}
& (INVmixLSB(ISB(INVmixLSB(x_{12}, x_{13}, k_{1012}, k_{1013})), \\
& ISB((INVmixMSB(x_{12}, x_{13}, k_{1012}, k_{1013})), k_{912}, k_{913})) \oplus \\
& (INVmixLSB(ISB(INVmixLSB(x_{12} + D_{131}, x_{13} + D_{132}, k_{1012}, k_{1013}))), \\
& ISB((INVmixMSB(x_{12} + D_{131}, x_{13} + D_{132}, k_{1012}, k_{1013}))), k_{912}, k_{913}) = \\
& (INVmix(ISB(INVmix(x_9, k_{109})), k_{99})) \oplus \\
& (INVmix(ISB(INVmix(x_9 + D_{14}, k_{109}))), k_{99}))
\end{aligned} \tag{10}
$$

$$
\begin{aligned}
& (INVmixLSB(ISB(INVmixLSB(x_6, x_7, k_{106}, k_{107})), \\
& ISB((INVmixMSB(x_6, x_7, k_{106}, k_{107})), k_{914}, k_{915})) \oplus \\
& (INVmixLSB(ISB(INVmixLSB(x_6 + D_{151}, x_7 + D_{152}, k_{106}, k_{107}))), \\
& ISB((INVmixMSB(x_6 + D_{151}, x_7 + D_{152}, k_{106}, k_{107}))), k_{914}, k_{915}) = \\
& 3[(INVmix(ISB(INVmix(x_9, k_{109})), k_{99})) \oplus \\
& (INVmix(ISB(INVmix(x_9 + D_{14}, k_{109}))), k_{99}))]
\end{aligned} \tag{11}
$$

$$
\begin{aligned}
& (INVmix(ISB(INVmix(x_3, k_{103})), k_{915})) \oplus \\
& (INVmix(ISB(INVmix(x_3 + D_{16}, k_{103}))), k_{915})) = \\
& 2[(INVmixLSB(ISB(INVmixLSB(x_{12}, x_{13}, k_{1012}, k_{1013})), \\
& ISB((INVmixMSB(x_{12}, x_{13}, k_{1012}, k_{1013})), k_{912}, k_{913})) \oplus \\
& (INVmixLSB(ISB(INVmixLSB(x_{12} + D_{131}, x_{13} + D_{132}, k_{1012}, k_{1013}))), \\
& ISB((INVmixMSB(x_{12} + D_{131}, x_{13} + D_{132}, k_{1012}, k_{1013}))), k_{912}, k_{913})]
\end{aligned} \tag{12}
$$

10th round keys k_{103}, k_{106}, k_{107}, k_{109}, k_{1012}, and k_{1013} and 9th round keys k_{912}, k_{913}, k_{914}, and k_{915} are obtained from Eqs. (10), (11), and (12).

3.1 A Working Example

Assume $PT1$ is a given plaintext

$$PT1 = \begin{pmatrix} 75 & 52 & 44 & 34 \\ 17 & 76 & 60 & 28 \\ 37 & 57 & a5 & fc \\ 84 & 6d & a1 & 50 \end{pmatrix}$$

The cipher key K_0 is as follows

$$K_0 = \begin{pmatrix} c7 & bd & d7 & be \\ 9b & d9 & 9b & 9c \\ da & cd & 6c & fa \\ bc & 28 & f8 & 9c \end{pmatrix}$$

The 9th round key obtained by employing AES key expansion algorithm is as follows

$$K_9 = \begin{pmatrix} af & 35 & 24 & 90 \\ f0 & fa & 45 & 99 \\ a7 & 87 & 34 & 33 \\ d8 & 17 & be & 59 \end{pmatrix}$$

and the 10th round key is as follows

$$K_{10} = \begin{pmatrix} 77 & 42 & 66 & f6 \\ 33 & c9 & 8c & 15 \\ 6c & eb & df & ec \\ b8 & af & 11 & 48 \end{pmatrix}$$

$$CT1 = \begin{pmatrix} 9 & e4 & 2c & ce \\ 6c & a6 & 30 & 82 \\ 74 & 5 & 93 & 36 \\ 13 & 3a & 0e & b4 \end{pmatrix}$$

The corresponding faulty ciphertext after injecting fault at 00th byte of 9th round is as follows

$$CT1' = \begin{pmatrix} \mathbf{e5} \ e4 \ 2c \ ce \\ \mathbf{93} \ a6 \ 30 \ \mathbf{d3} \\ 74 \ 5 \ \mathbf{6d} \ 36 \\ 13 \ \mathbf{7e} \ \mathbf{f1} \ b4 \end{pmatrix}$$

The bolded bytes show how the fault has been propagated to the ciphertext.

Let another plaintext be

$$PT2 = \begin{pmatrix} 99 \ 63 \ 73 \ 7e \\ 6a \ 4 \ 2e \ 1c \\ 52 \ 90 \ ed \ ee \\ 28 \ c8 \ 75 \ a \end{pmatrix}$$

The ciphertext for the key K_0 is as follows

$$CT2 = \begin{pmatrix} 3c \ ec \ 44 \ f7 \\ 91 \ 3a \ 82 \ 2d \\ ef \ 70 \ 2a \ 1b \\ 67 \ d9 \ da \ cf \end{pmatrix}$$

The corresponding faulty ciphertext after injecting fault at 00th byte is as follows

$$CT2' = \begin{pmatrix} \mathbf{98} \ ec \ 44 \ f7 \\ \mathbf{93} \ 3a \ 82 \ \mathbf{da} \\ ef \ 70 \ \mathbf{fd} \ 1b \\ 67 \ \mathbf{1e} \ \mathbf{25} \ cf \end{pmatrix}$$

Employing Eqs. (1), (2), and (3), it is giving exact 10th round keys k_{100}, k_{101}, k_{107}, k_{1010}, k_{1011}, k_{1013}.

3.2 Comparison with Existing Works

In this section, a comparison is provided in terms of fault model, fault location, number of faulty encryptions, and computational complexity of our work and with the works reported in [5–9, 12]. Existing related works either based on byte-level or bit-level fault model. Our work is based on byte-level fault model. From Table 1, it is observed that in the proposed architecture, 24 faulty–fault-free ciphertext pairs are necessary to mount fault attack by injecting fault at the input of 9th round. Also fault attack complexity in the proposed scheme is relatively higher than that of AES [12]. Fault attack on AES [12] requires minimum 2 faulty–fault-free ciphertext pairs with complexity 2^{32}. Fault attack on MDS-AES [9] needs 2 faulty ciphertext pairs with

Table 1 Comparison of existing fault attack on AES with our proposed SPN-type architecture accomplishing properties of the encryption function

Reference	Fault model	Algorithm	Fault location	No. of faulty encryp- tions	Complexity
Blomer and Seifert [7]	Force 1 bit to 0	AES	Chosen	128	
Blomer and Seifert [7]	Implementation dependent	AES	Chosen	256	
Giraud [6]	Switch 1 bit	AES	Any bit of chosen bytes	50	
Giraud [6]	Disturb 1 byte	AES	Anywhere among 4 bytes	250	
Dusart et al. [5]	Disturb 1 byte	AES	Anywhere between last two MixColumn	40	
Piret and Quisquater [8]	Disturb 1 byte	AES	Anywhere between 7th round and 7th round MixColumn	2	
Mukhopadhyay [12]	Disturb 1 byte	AES	Anywhere between 7th round	2	2^{32}
Das and Bhaumik [9]	Disturb 1 byte	MDS-AES	Input of 9th round and 7th round MixColumn	2	2^{16}
This paper	Disturb 1 byte	New SPN architecture	Input at 9th round	24	2^{53}

brute-force search of complexity 2^{16}, whereas to mount fault attack on the proposed SPN-type architecture needs 24 faulty–fault-free ciphertext pairs with computation complexity 2^{53}.

4 Conclusion

This paper presents a new SPN-type architecture which can improve the security of block cipher against fault attack. In this paper, instead of linear byte-wise round key mixing a 16-bit nonlinear round key mixing is used and its security has been analyzed against fault attack. The proposed new SPN architecture is giving greater security than original AES. When fault is injected at the input of 9th round SubByte operation, computation requires 24 pairs of faulty and fault-free ciphertexts. For retrieving 10th round key of 128 bits, it costs the computation complexity of $2^5 3$.

References

1. D. Boneh, R.A. DeMillo, R.J. Lipton, On the importance of eliminating errors in cryptographic computations. J. Cryptol. **12**, 241–246 (2001)
2. D. Boneh, R.A. DeMillo, R.J. Lipton, On the importance of checking cryptographic protocols for faults, in *EUROCRYPT 1997*. LNCS, vol. 1233 (1997), pp. 37–51
3. E. Biham, A. Shamir, *Differential Fault Analysis of Secret Key Cryptosystems*. CRYPTO 1997, LNCS, vol. 1294 (1997), pp. 513–525
4. J. Daemen, V. Rijmen, *The Design of Rijndael* (Springer, Heidelberg, 2002)
5. P. Dusart, G. Letourneux, O. Vivolo, *Differential Fault Analysis on A.E.S.* (2002). http://eprint.iacr.org/2003/010
6. C. Giraud, DFA on AES, *Cryptology ePrint Archive*, Report 2003/008
7. J. Blomer, J.P. Seifert, *Fault Based Cryptanalysis of the Advanced Encryption Standard (AES)*, ed. by R.N. Wright. FC 2003, LNCS, vol. 2742 (2003), pp. 162–181
8. G. Piret, J.J. Quisquater, *A Differential Fault Attack Technique Against SPN Structures, with Application to the AES and Khazad*. CHES 2003, LNCS, vol. 2779 (2003), pp. 77–88
9. S. Das, J. Bhaumik, A fault based attack on MDS-AES. Int. J. Netw. Secur. **16**(3), 193–198 (2014)
10. S. Ali, X. Guo, R. Karri, D. Mukhopadhyay, Fault attacks on AES and their countermeasures, in *Secure System Design and Trustable Computing*, Book Part: Part I (2016), pp. 163–208. https://doi.org/10.1007/978-3-319-14971-4-5
11. J. Bhaumik, D. Roy Chowdhury, NMIX: an ideal candidate for key mixing, in *Proceedings of the International Conference on Security and Cryptography* (2009), pp. 285–288
12. D. Mukhopadhyay, *An Improved Fault Based Attack of the Advanced Encryption Standard*. AFRICACRYPT, LNCS, vol. 5580 (2009), pp. 421–434

High-Capacity Reversible Data Hiding Scheme Using Dual Color Image Through (7, 4) Hamming Code

Ananya Banerjee and Biswapati Jana

Abstract Achievement of high-capacity data hiding with good visual quality is an important research issue in the field of steganography. In this paper, we have introduced a new dual color image-based reversible data hiding scheme through (7, 4) Hamming code and shared a secret key. We partitioned the color image into (3 × 3) pixel blocks and then decomposed into three basic color blocks. Again each color blocks are sliced up to 4-bit plane starting from LSB plane. Now, a segment of 3-bit secret data is embedded within each bit plane depending on a syndrome calculated using (7, 4) Hamming code. As a result, 36-bit secret data can be embedded within a (3 × 3) pixel block. Therefore, we achieve a high payload good visual quality stego compared with existing schemes. Dual stego images are generated to achieve reversibility in (7, 4) Hamming code-based data hiding. The secret data and the original cover image both are successfully retrieved and recovered from dual stego images.

Keywords Steganography · Hamming code · Least significant bit (LSB)
Bit plane · Reversible data hiding · Dual image

1 Introduction

Steganography is the art and science of hidden data communication. It is useful in many application areas to solve the problem of ownership identification, copyright protection, authentication, verification, and more. The main aim of data hiding schemes is to ensure extraction of secret data and recovery of the original object from stego media. The data embedding in color image is considered to be more

A. Banerjee (✉) · B. Jana
Department of Computer Science, Vidyasagar University,
Midnapore 721101, West Bengal, India
e-mail: anaanya.2011@gmail.com

B. Jana
e-mail: biswapatijana@gmail.com

© Springer Nature Singapore Pte Ltd. 2017
J. Bhaumik et al. (eds.), *Communication, Devices, and Computing*, Lecture Notes in Electrical Engineering 470, https://doi.org/10.1007/978-981-10-8585-7_12

unsuspicious and secured. An error-correcting code could not only detect that errors have occurred but also locate the error positions. Hamming code is a linear error-correcting code that can detect and correct single-bit errors. The $(n, n - k)$ Hamming code uses n cover bits to transmit $n - k$ message bits, and the other k bits used for error correcting purpose are called parity check bits, where $n = 2k - 1$ on the binary file. The (7, 4) Hamming code is now taken as an example to demonstrate how Hamming code corrects an error bit.

Data hiding is mainly classified into two broad categories: irreversible [1, 2] and reversible [3–7]. Ni et al. [3] introduced reversible data hiding based on histogram shifting. Data hiding through matrix coding has been introduced by Crandall [8], and Westfeld [9] implemented F5 algorithm based on matrix encoding using hamming code. Kim and Shin [2] suggested a data embedding procedure for halftone image. The scheme provides good capacity but poor visual quality. Based on matrix encoding, 'Hamming +1' method has been developed by Zhang et al. [10]. The embedding capacity is increased in 'Hamming +1' scheme by $\frac{k+1}{2^k}$ bpp. Chang and Chou [11] suggested and improved data embedding procedure using Hamming code which can hide (k + 1) bits of a message in 2^k pixels with at most one change. Tian [6] developed an RDH scheme using difference expansion. The disadvantage of the scheme is overflow and underflow problems.

Recently dual image-based RDH becomes very important for medical image processing and military communications. Information can be embedded into an image which contains ownership identification, authentication, and copyright protection. Chang et al. [12] proposed a RDH scheme using two images. Lee et al. [13] developed an RDH scheme using dual stego images, in which only one pixel value needs to be modified by at most either plus or minus 1 for carrying 2-bit secret data. Lee and Huang [14] developed a dual image-based RDH method that overcomes the drawback of the previous method. Jana et al. [15] proposed an RDH scheme using dual image with 53 dB PSNR and payload 0.14 bpp.

Here, we have proposed an improved dual image-based data hiding scheme using (7, 4) Hamming code for color images. We have divided RGB color pixels in bit plane starting from LSB to LSB-3 (up to 4-bit plane) into (3 × 3) blocks and then apply (7, 4) Hamming code-based data hiding scheme. As a result, we increase the payload up to 4 bpp. In this scheme, the receiver can successfully extract secret data and recover the original cover image from dual stego images.

The rest of the paper is organized as follows. Section 2 presents the proposed method and numerical illustration. In Section 3, we discuss experimental results and analyze them. In Section 4, we introduce security analysis. Section 5 provides the conclusion.

2 Proposed Method

Here, I is considered as the cover image of size (M × N), and I' is the marked image with data D = {d_1, ..., d_X} which is embedded, where $d_i \in \{0, 1\}$, $1 \le i \le X$. Here, H is a parity check matrix of the Hamming code and can be expressed as

$$H = \begin{vmatrix} 0\ 0\ 0\ 1\ 1\ 1\ 1\ 0\ 0 \\ 0\ 0\ 1\ 0\ 1\ 1\ 0\ 1\ 0 \\ 0\ 0\ 1\ 1\ 0\ 1\ 0\ 0\ 1 \end{vmatrix}.$$

Before embedding, we have considered 36-bit shared secret key k_1 to encrypt secret data bits using symmetric key encryption. We have taken pixel block of size (3 × 3), and 4-bit plane of each pixel is used to embed secret data, as a result (3 × 3) × 4 = 36 bits of data (D_1) are embedded in one iteration. To enhance additional security measures, instead of choosing the cover image pixel block serially, we use Pseudo Random Number Generator [16] (PRNG) function with a secret predefined seed k_2 (which is only known to the sender and the receiver) to determine the next available block for embedding. Since this seed will be known to the sender and receiver only, the generated unique pattern of pixel block selection can be used in embedding and extraction process securely. The data embedding procedure is enlisted in Algorithm 1, and the data extraction procedure is depicted in Algorithm 2.

Algorithm 1: Data embedding process
Input: The color cover image I (M × N), secret data bits D, Hamming matrix H, secret key k_1, and seed value k_2.
Output: Two stego images I' and I'' of size (M × N).
Step 1: Collect a random sequence of pixel blocks of size 3 × 3 from $I_{M \times N}$ using PRNG (k_2). Say the pixel blocks are X_1, X_2, ..., X_{MN}.
Step 2: Convert X_i into 3 separate RGB color blocks X_{iR}, X_{iG}, X_{iB}.
Step 3: Convert each X_i's into binary form.
Step 4: Perform bit plane slicing of each X_i's up to 4-bit plane starting from LSB, that is, $X_{iR(LSB)}$, $X_{iR(LSB-1)}$, $X_{iR(LSB-2)}$, $X_{iR(LSB-3)}$.
Step 5: Perform $D_1' = (D_1 \oplus k_1)$; k_1 =36 bit length and D_1 is also same length.
Step 6: Select 3 bits of data from secret data stream, i.e. d_i = {d_1, d_2, d_3} from D_1' where $d_i \in \{0, 1\}$.
Step 7: Convert $X_{iR(LSB)}$ into a 1D matrix.
Step 8.1: Take c = $X_{iR(LSB)}$ and calculate the syndrome $S_1 = (H \times (c)^T)^T$.
Step 8.2: Calculate $S_2 = (d_i \oplus S_1)$; if S_2=0, no change, otherwise flip a bit at the positional value of S_2 and generate H'.
Step 8.3: Compute $S_3 = (H' \oplus c)$ and store the data.
Step 8.4: Replace the matrix(c) with S_3 and update $X_{iR(LSB)}$ for first stego image.
Step 9: Take c = $X_{iR(LSB)}$ and perform c'= ($d_i \oplus c$).
Step 9.2: Compute 1's complement of c' starting from LSB up to (LSB-2) position.

Step 9.3: Replace the matrix(c) with c' and update $X_{iR(LSB)}$ for second stego image.

Step 10: Repeat **Step 4** to **9** using $X_{iR(LSB-1)}$, $X_{iR(LSB-2)}$ and $X_{iR(LSB-3)}$.

Step 11: Repeat **Step 5** to **11** to embed secret data on X_{iG} and X_{iB} color blocks.

Step 12: Repeat **Step 2** to **12** to embed secret data on each and every random sequence of (X_i's) of pixel blocks.

Step 13: Finally after combining each stego block, we get two stego images (I') and (I") of size (M × N).

Step 14: End.

Algorithm 2: Data extraction process

Input: Two stego images I' and I" of size (M × N), Hamming matrix H, secret key k_1, and seed value k_2.

Output: Original secret message D, original cover image I.

Step 1: Use PRNG with predetermined seed k_2 to determine the stego pixel of random sequence X'_i of size [3 × 3] from stego image I' and I".

Step 2: Separate RGB components into X'_{iR}, X'_{iG}, X'_{iB}.

Step 3: Convert into binary form of each X'_{iR}, X'_{iG}, X'_{iB}.

Step 4: Perform 4-bit plane slicing of each X'_i 's starting from LSB that is $X'_{iR(LSB)}$, $X'_{iR(LSB-1)}$, $X'_{iR(LSB-2)}$, $X'_{iR(LSB-3)}$.

Step 5: Convert $X_{iR(LSB)}$ into 1D matrix.

Step 6.1: From the stego image I', Take $c' = X'_{iR(LSB)}$ and calculate the syndrome $S' = (H \times (c')^T)^T$.

Step 6.2: Concatenate syndrome S' with data unit of D' that is $D' = D' \| (S')$.

Step 6.3: Compute $D_i = D' \oplus k_1$.

Step 6.4: Concatenate D_i's, we get original secret message D.

Step 7.1: From the stego image I", compute 1's complement of c' starting from LSB up to (LSB-2) position.

Step 7.2: Perform $c' = (D_i' \oplus c)$ which is obtained from **Step 6.4** from the first stego image.

Step 8: Repeat **Step 4** to **7** using $X_{iR(LSB-1)}$, $X_{iR(LSB-2)}$ and $X_{iR(LSB-3)}$.

Step 9: Repeat **Step 5** to **8** to embed secret data on X_{iG} and X_{iB} color blocks.

Step 10: Repeat **Step 2** to **9** to embed secret data on each and every random sequence of (X_i's) of pixel blocks.

Step 11: Finally after combining each stego block, we get original cover image (I) of size (M × N).

Step 12: End.

2.1 Numerical Illustration

Example 2.1.1: Data Embedding

1. Let I is a color image block with (3 × 3) pixel. D = {d_1, d_2, ..., d_{36}} = {0, 1, 0, 1, 0, 1, 0, 0, 1, 1, 0, 1, 1, 1, 0, 0, 0, 1, 1, 0, 1, 0, 0, 1, 1,0, 1, 0, 1, 0, 0, 0, 1, 0, 1,

0}. $k_1 = \{0, 0, 1, 0, 1, 1, 0, 0, 1, 0, 0, 0, 1, 0, 1, 0, 0, 0, 1, 1, 1, 0, 0, 1, 0, 0, 1, 0, 1, 0, 1, 0, 1, 0, 1, 0\}$ and $ER = 36/(3 \times 3) = 4$ bpp, and the cover image pixels are as follows:

$$I_{3 \times 3} = \begin{vmatrix} 9277330 & 9276816 & 9277072 \\ 9276816 & 9343124 & 9343381 \\ 9211793 & 9409173 & 9409173 \end{vmatrix}$$

and $D' = D \oplus k_1 = \{0, 1, 1, 1, 1, 0, 0, 0, 0, 1, 0, 1, 0, 1, 1, 0, 0, 1, 0, 1, 0, 0, 0, 0, 1, 0, 0, 0, 0, 0, 1, 0, 0, 0, 0, 0\}$ (Fig. 1).

2. Divided into 3 RGB image pixel blocks shown below.

$$R = \begin{vmatrix} 141 & 141 & 141 \\ 141 & 142 & 142 \\ 140 & 143 & 143 \end{vmatrix} \quad G = \begin{vmatrix} 143 & 141 & 142 \\ 141 & 144 & 145 \\ 143 & 146 & 147 \end{vmatrix} \quad B = \begin{vmatrix} 147 & 145 & 145 \\ 145 & 149 & 150 \\ 146 & 150 & 151 \end{vmatrix}$$

3. Take Red image pixel block and transform into binary number matrix.

$$R = \begin{vmatrix} 10001101 & 10001101 & 10001101 \\ 10001101 & 10001110 & 10001110 \\ 10001100 & 10001111 & 10001111 \end{vmatrix}$$

4. Divide it into 4-bit plane matrices starting from LSB.

$$R_{LSB} = \begin{vmatrix} 1 & 1 & 1 \\ 1 & 0 & 0 \\ 0 & 1 & 1 \end{vmatrix} \quad R_{LSB-1} = \begin{vmatrix} 0 & 0 & 0 \\ 0 & 1 & 1 \\ 0 & 1 & 1 \end{vmatrix}$$

$$R_{LSB-2} = \begin{vmatrix} 1 & 1 & 1 \\ 1 & 1 & 1 \\ 1 & 1 & 1 \end{vmatrix} \quad R_{LSB-3} = \begin{vmatrix} 1 & 1 & 1 \\ 1 & 1 & 1 \\ 1 & 1 & 1 \end{vmatrix}$$

5. Read the LSB matrix and form a 1D matrix. $c = [1\ 1\ 1\ 1\ 0\ 0\ 0\ 1\ 1]$
6. XOR data with 1D matrix for second stego image $c' = c \oplus D' = [1\ 1\ 1\ 1\ 0\ 0\ 0\ 1\ 1] \oplus [0\ 0\ 0\ 0\ 0\ 0\ 0\ 1\ 1] = [1\ 1\ 1\ 1\ 1\ 0\ 0\ 0\ 0]$
7. 1's complement up to last 3 bits then we will get $c' = [1\ 1\ 1\ 1\ 0\ 0\ 1\ 1\ 1]$. For the first stego image, go to step 8. Go to step 11 and generate the second stego image.

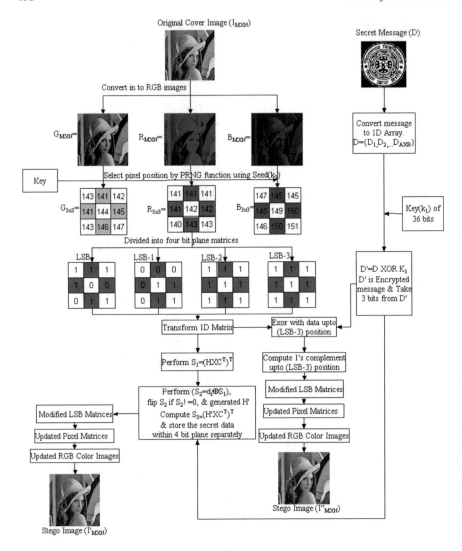

Fig. 1 Pictorial diagram of the proposed data hiding scheme

8. Calculate the syndrome $S_1 = H \times (c)^T = \begin{vmatrix} 0\ 0\ 0\ 1\ 1\ 1\ 1\ 0\ 0 \\ 0\ 0\ 1\ 0\ 1\ 1\ 0\ 1\ 0 \\ 0\ 0\ 1\ 1\ 0\ 1\ 0\ 0\ 1 \end{vmatrix} \times$

$\begin{vmatrix} 1\ 1\ 1\ 1\ 1\ 0\ 0\ 0\ 1\ 1 \end{vmatrix}^T = \begin{vmatrix} 1 \\ 0 \\ 1 \end{vmatrix}$

9. Transpose the syndrome and XOR with the secret data bit, i.e., $[1\ 0\ 1] \oplus [0\ 1\ 1] = [1\ 1\ 0]$ which is match with the fifth column of Hamming matrix.

10. Generate the code H' = [0 0 0 0 1 0 0 0 0] and XOR with the original code c.
$S_3 = [1\ 1\ 1\ 1\ 0\ 0\ 1\ 1] \oplus [0\ 0\ 0\ 0\ 1\ 0\ 0\ 0\ 0] = [1\ 1\ 1\ 1\ 1\ 0\ 0\ 1\ 1].$

11. Transform into a new LSB matrix.

$$R'_{LSB} = \begin{vmatrix} 1 & 1 & 1 \\ 1 & 1 & 0 \\ 0 & 1 & 1 \end{vmatrix}$$

12. Similarly compute the LSB-1, LSB-2, and LSB-3 matrices as follows:

$$R'_{LSB-1} = \begin{vmatrix} 0 & 0 & 0 \\ 0 & 1 & 0 \\ 0 & 1 & 1 \end{vmatrix} \quad R'_{LSB-2} = \begin{vmatrix} 1 & 1 & 1 \\ 1 & 1 & 1 \\ 1 & 1 & 0 \end{vmatrix} \quad R'_{LSB-3} = \begin{vmatrix} 1 & 1 & 1 \\ 0 & 1 & 1 \\ 1 & 1 & 1 \end{vmatrix}$$

13. Update all four modified binary matrices to their corresponding position in original Red pixel matrix.

$$R'_{3\times3} = \begin{vmatrix} 10001101 & 10001101 & 10001101 \\ 10000101 & 10001111 & 10001100 \\ 10001100 & 10001111 & 10001011 \end{vmatrix} = \begin{vmatrix} 141 & 141 & 141 \\ 133 & 143 & 140 \\ 140 & 143 & 139 \end{vmatrix}$$

14. Similarly, get updated Green and Blue pixel matrices.

$$G'_{3\times3} = \begin{vmatrix} 143 & 141 & 134 \\ 143 & 144 & 149 \\ 143 & 146 & 146 \end{vmatrix} \quad B'_{3\times3} = \begin{vmatrix} 147 & 145 & 149 \\ 145 & 149 & 150 \\ 147 & 156 & 151 \end{vmatrix}$$

15. Finally, the first stego image block will be $I'_{3\times3} =$
$$\begin{vmatrix} 9277330 & 9276816 & 9275028 \\ 8753040 & 9408660 & 9213333 \\ 9211794 & 9409179 & 9409179 \end{vmatrix}, \text{ and second stego image block } I''_{3\times3} =$$
$$\begin{vmatrix} 9277310 & 9278816 & 9276728 \\ 8753340 & 9437661 & 9217853 \\ 9264995 & 9402109 & 9469459 \end{vmatrix}$$

Example 2.1.2: Data Extraction

1. The marked image size I' of size 3×3 is shown below.

$$I' = \begin{vmatrix} 9277330 & 9276816 & 9275028 \\ 8753040 & 9408660 & 9213333 \\ 9211794 & 9409179 & 9409179 \end{vmatrix}$$

2. Divided into 3 RGB image pixel blocks.

$$R = \begin{vmatrix} 141 & 141 & 141 \\ 133 & 143 & 140 \\ 140 & 143 & 139 \end{vmatrix} \quad G = \begin{vmatrix} 143 & 141 & 134 \\ 143 & 144 & 149 \\ 143 & 146 & 146 \end{vmatrix} \quad B = \begin{vmatrix} 147 & 145 & 149 \\ 145 & 149 & 150 \\ 147 & 156 & 151 \end{vmatrix}$$

3. Take Red image pixel block and transform into binary numbers.

$$\begin{vmatrix} 10001101 & 10001101 & 10001101 \\ 10000101 & 10001111 & 10001100 \\ 10001100 & 10001111 & 10001011 \end{vmatrix}$$

4. Divide it into 4-bit plane matrices starting from LSB.

$$R_{LSB} = \begin{vmatrix} 1 & 1 & 1 \\ 1 & 1 & 0 \\ 0 & 1 & 1 \end{vmatrix} \quad R_{LSB-1} = \begin{vmatrix} 0 & 0 & 0 \\ 0 & 1 & 0 \\ 0 & 1 & 1 \end{vmatrix}$$

$$R_{LSB-2} = \begin{vmatrix} 1 & 1 & 1 \\ 1 & 1 & 1 \\ 1 & 1 & 0 \end{vmatrix} \quad R_{LSB-3} = \begin{vmatrix} 1 & 1 & 1 \\ 0 & 1 & 1 \\ 1 & 1 & 1 \end{vmatrix}$$

5. Read LSB matrix and form a 1D matrix. $c = [1\ 1\ 1\ 1\ 1\ 0\ 0\ 1\ 1]$
6. Calculate the syndrome.

$$S_1 = H \times (c)^T = \begin{vmatrix} 0 & 0 & 0 & 1 & 1 & 1 & 1 & 0 & 0 \\ 0 & 0 & 1 & 0 & 1 & 1 & 0 & 1 & 0 \\ 0 & 0 & 1 & 1 & 0 & 1 & 0 & 0 & 1 \end{vmatrix} \times \begin{vmatrix} 1 & 1 & 1 & 1 & 1 & 0 & 0 & 1 & 1 \end{vmatrix}^T = \begin{bmatrix} 0 \\ 1 \\ 1 \end{bmatrix}$$

7. Transpose the syndrome to get secret data bits $d = [0\ 1\ 1]$
8. Repeat the above steps until we do not get the secret data bits. Concatenate all the data bits to get the data, i.e., D' = {0, 1, 1, 1, 0, 1, 0, 0, 1, 1, 0, 1, 1, 1, 0, 0, 0, 1, 1, 0, 1, 0, 0, 1, 1, 0, 1, 0, 1, 0, 0, 0, 1, 0, 1, 0}.
9. From the second stego, we will do again 1's complement and get
   $c' = [1\ 1\ 1\ 1\ 0\ 0\ 0\ 0\ 0]$
10. Next XORed with data which is generated from the first stego image then we will get the original pixel matrix $c' = c \oplus D' = [1\ 1\ 1\ 1\ 0\ 0\ 0\ 0\ 0] \oplus [0\ 0\ 0\ 0\ 0\ 0\ 0\ 1\ 1] = [1\ 1\ 1\ 1\ 0\ 0\ 0\ 1\ 1]$. After combining all cover matrixes, we will get the original cover image.
11. XOR the modified secret data with secret key k_1 to get the original secret data bits, i.e., D = {0, 1, 0, 1, 0, 1, 0, 0, 1, 1, 0, 1, 1, 1, 0, 0, 0, 1, 1, 0, 1, 0, 0, 1, 1, 0, 1, 0, 1, 0, 0, 0, 1, 0, 1, 0}.

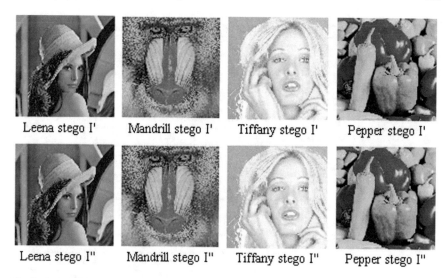

Fig. 2 Stego images of size (512 × 512)

3 Experimental Result and Comparison

The scheme is implemented using NetBeans IDE 8.0 on standard color images to measure the performance. The standard cover images are collected from Image database of SIPI [17]. The quality of the stego images is measured using Mean Square Error (MSE) and peak signal-to-noise ratio (PSNR). Figure 2 shows the stego image when ER = 4 bpp.

$$MSE = \frac{1}{M \times N} \sum_{i=1}^{M} \sum_{j=1}^{N} [I(i, j) - I'(i, j)]^2 \tag{1}$$

$$PSNR = 10 \log_{10} \frac{255^2}{MSE} (dB) \tag{2}$$

Following Table 1 represents the comparison of PSNR values of stego images generated by different methods with varying payload, and we achieved better PSNR value every time compared to existing methods. The time complexity of our algorithm is 0(4n) where n represents the number of pixels. The execution time of the data embedding algorithm required 10 s when we use Pentium 4 machine with 2 GB RAM to embed 780,300 bits within a 512 × 512 color cover image.

Table 1 Comparison of image quality between 9 dual image-based methods

Method	Measure	Lenna	Peppers	Mandrill
Chang et al. [12]	PSNR (1)	45.12	45.14	45.11
	PSNR (2)	45.13	45.15	45.13
	PSNR (Avg)	45.13	45.15	45.12
	Capacity	524,288	523,356	524,148
Chang et al. [18]	PSNR (1)	48.13	48.11	48.14
	PSNR (2)	48.14	48.14	48.13
	PSNR (Avg)	48.14	48.13	48.14
	Capacity	524,288	524,288	524,288
Lee et al. [13]	PSNR (1)	51.14	51.14	51.14
	PSNR (2)	54.16	54.17	54.14
	PSNR (Avg)	52.65	52.66	52.64
	Capacity	392,880	392,796	393,486
Lee and Huang [14]	PSNR (1)	49.76	49.75	49.77
	PSNR (2)	49.56	49.56	49.56
	PSNR (Avg)	49.66	49.66	49.67
	Capacity	560,801	560,572	560,686
Chang et al. [19]	PSNR (1)	39.89	39.94	39.91
	PSNR (2)	39.89	39.94	39.91
	PSNR (Avg)	39.89	39.94	39.91
	Capacity	802,895	799,684	802,524
Qin et al. [20]	PSNR (1)	52.11	51.25	52.04
	PSNR (2)	41.58	41.52	41.56
	PSNR (Avg)	46.85	46.39	46.80
	Capacity	557,052	557,245	557,096
Lu et al. [21]	PSNR (1)	49.20	49.19	49.21
	PSNR (2)	49.21	49.21	49.2
	PSNR (Avg)	49.21	49.20	49.21
	Capacity	524,288	524,192	524,204
Jana et al. [15]	PSNR (1)	52.71	52.67	52.70
	PSNR (2)	52.81	52.72	52.76
	PSNR (Avg)	52.76	52.69	52.73
	Capacity	74,898	74,898	74,898
Proposed method	PSNR (1)	44.04	44.01	44.05
	PSNR (2)	51.15	51.16	51.15
	PSNR (Avg)	47.59	47.58	47.60
	Capacity	780,300	780,300	780,300

Table 2 Experimental results of average SSIM and SD for ER = 4 bpp

	Lenna	Baboon	Tiffany	Peppers	Jet	Sailboat	Splash
SSIM	0.9305	0.9767	0.9386	0.93022	0.9392	0.96248	0.9244
Standard deviation	0.0815	0.1110	0.1754	0.04630	0.0768	0.05825	0.02812

Table 3 Average RS Analysis of two stego image with ER = 4 bpp

	Lenna	Woodland Hills	House	Peppers	Jet	Sailboat
R_m	18,137	17,288	18,109	17,427	18,294	17,508
R_{m1}	19,056	18,152	19,237	18,151	19,637	18,123
S_m	14,555	15,772	14,805	15,396	14,427	15,509
S_{m1}	13,868	15,006	13,983	14,825	13,362	15,053
RS value	0.0491	0.0493	0.0592	0.0395	0.0736	0.0324

4 Security Analysis

Security analysis is an important measure of data hiding process. In this paper, we have used two security levels to enhance our proposition from a security perspective. First, we consider a 36-bit secret key and encrypt the secret data bits using symmetric key encryption. As it is only known to the sender and receiver, the third party will not be able to decrypt it without knowing the secret key. In the second level of security, we have taken a secret seed which also known to the receiver and sender only. Using this seed, we generate a sequence of unique numbers with the help of PRNG function. We have taken the cover image pixel blocks according to the generated numbers. So without knowing this seed, no one will be able to predict the number sequence (Table 2).

We also verified our algorithm against some standard measurement like RS Analysis, SSIM, standard deviation, and histogram analysis. The Structural Similarity (SSIM) index is a method for measuring the similarity between two images. The SSIM measures the perceptual difference between the original image and the stego image. From Table 2, it is observed that the SSIM values of all test images are close to 1. Standard deviation is used to measure the amount of variation between original and stego images. Here, we achieve SD close to zero means that the stego images and cover image are similar in nature. We also analyze our stego image through RS Analysis [22]. From Table 3, it is shown that the values of R_m and R_{m1}, S_m and S_{m1} are almost equal and the ratio of R and S lies around 0.05, which is very small, so that we can conclude that our proposed scheme is secure against RS attack. Figure 3 represents the histogram of the original cover image and the stego image. It is shown that the shape of the histogram almost remains same after embedding high-capacity secret data which is robust against histogram attack.

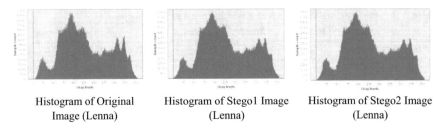

| Histogram of Original Image (Lenna) | Histogram of Stego1 Image (Lenna) | Histogram of Stego2 Image (Lenna) |

Fig. 3 Histogram of original and stego image

5 Conclusion

In this paper, we introduced a novel secure data hiding scheme using Hamming code for RGB color image. Bit plane slicing of each RGB color cover image block is also introduced to increase data hiding capacity over grayscale image. So the data embedding rate is raised up to 4 bpp which is greater than other existing schemes. We have tested our stego image with RS Analysis, histogram analysis, SSIM, SD method and observed that the proposed scheme is preferable for data embedding where visual quality and security constraint need to be maintained for high payload. In future, the scheme has been extended to enhance security, capacity, and quality in the different domain for video-based steganography.

References

1. C. Kim, C.N. Yang, Improving data hiding capacity based on hamming code, in *Frontier and Innovation in Future Computing and Communications* (Springer, Netherlands, 2014), pp. 697–706
2. H.J. Kim, C. Kim, Y. Choi, S. Wang, X. Zhang, Improved modification direction methods. Comput. Math Appl. **60**(2), 319–325 (2010)
3. Z. Ni, Y.Q. Shi, N. Ansari, W. Su, Reversible data hiding. IEEE Trans. Circuits Syst. Video Technol. **16**(3), 354–362 (2006)
4. C.F. Lee, H.L. Chen, Adjustable prediction-based reversible data hiding. Digit. Signal Process. **22**(6), 941–953 (2012)
5. C.F. Lee, H.L. Chen, H.K. Tso, Embedding capacity raising in reversible data hiding based on prediction of difference expansion. J. Syst. Softw. **83**(10), 1864–1872 (2010)
6. J. Tian, Reversible data embedding using a difference expansion. IEEE Trans. Circuits Syst. Video Technol. **13**(8), 890–896 (2003)
7. H.W. Tseng, C.P. Hsieh, Prediction-based reversible data hiding. Inf. Sci. **179**(14), 2460–2469 (2009)
8. R. Crandall, Some notes on steganography, posted on steganography mailing list (1998), http://os.inf.tudresden.de/westfield/Crandall.pdf
9. A. Westfeld, F5—a steganographic algorithm, in *International Workshop on Information Hiding* (Springer, Berlin, 2001), pp. 289–302
10. W. Zhang, S. Wang, X. Zhang, Improving embedding efficiency of covering codes for applications in steganography. IEEE Commun. Lett. **11**(8) (2007)

11. C.C. Chang, Y.C. Chou, Using nearest covering codes to embed secret information in grayscale images, in *Proceedings of the 2nd International Conference on Ubiquitous Information Management and Communication* (ACM, 2008), pp. 315–320
12. C.C. Chang, T.D. Kieu, Y.C. Chou, Reversible data hiding scheme using two steganographic images, in *2007 IEEE Region 10 Conference, TENCON 2007* (IEEE, 2007), pp. 1–4
13. C.F. Lee, K.H. Wang, C.C. Chang, Y.L. Huang, A reversible data hiding scheme based on dual steganographic images, in *Proceedings of the 3rd International Conference on Ubiquitous Information Management and Communication* (ACM, 2009), pp. 228–237
14. C.F. Lee, Y.L. Huang, Reversible data hiding scheme based on dual stegano-images using orientation combinations. Telecommun. Syst. **52**(4), 2237–2247 (2013)
15. B. Jana, D. Giri, S.K. Mondal, Dual image based reversible data hiding scheme using (7, 4) hamming code. Multimedia Tools Appl. 1–23 (2016)
16. R.S. DeBellis, R.M. Sr Smith, P.C.C. Yeh, U.S. Patent No. 6,044,388. Washington, DC: U.S. Patent and Trademark Office (2000)
17. University of Southern California, The USC-SIPI Image Database (2015), http://sipi.usc.edu/database/database.php
18. C.C. Chang, Y.C. Chou, T.D. Kieu, Information hiding in dual images with reversibility, in *Proceedings of the 3rd International Conference on Multimedia and Ubiquitous Engineering* (2009), pp. 145–152
19. C.C. Chang, T.C. Lu, G. Horng, Y.H. Huang, Y.M. Hsu, A high payload data embedding scheme using dual stego-images with reversibility, in *Proceedings of the 3rd International Conference on Information, Communications and Signal Processing* (2013), pp. 1–5
20. C. Qin, C.C. Chang, T.J. Hsu, Reversible data hiding scheme based on exploiting modification direction with two steganographic images, Multimedia Tools Appl. 1–12 (2014)
21. T.C. Lu, C.Y. Tseng, J.H. Wu, Dual imaging-based reversible hiding technique using LSB matching. Signal Process. **108**, 77–89 (2015)
22. J. Fridrich, M. Goljan, R. Du, Invertible authentication, in *Photonics West 2001-Electronic Imaging, International Society for Optics and Photonics* (2001), pp. 197–208

A Study on the Effect of a Rectangular Slot on Miniaturization of Microstrip Patch Antenna

Sunandan Bhunia and Avisankar Roy

Abstract In this paper, a study on the effect of rectangular slot in a non-radiating edge of rectangular microstrip patch antenna on the miniaturization of the antenna has been presented. It has been investigated from parametric studies that frequency becomes lower due to the increase of effective electrical path length along the slot dimensions. The simulation has been done with the method of moment-based software IE3D, and the results have been discussed with equivalent circuit of slotted antenna. With the optimum dimension of the slot, almost 78% compactness has been found with respect to conventional patch antenna.

Keywords Rectangular microstrip patch · Rectangular slot · Compactness

1 Introduction

In recent years, the size of modern wireless communication devices became smaller which has led to the development of more compact antenna. Microstrip patch antennas are one of the most important parts of wireless communication devices due to some advantages like low-profile, lightweight, easy fabrication capabilities with RF circuits [1].

Several techniques may be used to design a compact microstrip patch antenna discussed in many literatures. Different types of slots may be incorporated on patch which increases the electrical length, and it causes lowering the frequency and results compactness. Slotted ground plane and monopole structures are also the different design procedures of compact microstrip patch antennas. Bhuina et al. [2, 3] have investigated on compactness and multi-frequency operation of microstrip patch antenna cutting some rectangular slots on the patch, and he has achieved almost 85% size reduction. Roy et al. [4, 5] have investigated on compactness of microstrip patch

S. Bhunia (✉)
Department of ECE, Central Institute of Technology, Kokrajhar, Assam, India
e-mail: snb.hit@gmail.com

A. Roy
Department of ECE, Haldia Institute of Technology, Haldia, East Medinipur, India

© Springer Nature Singapore Pte Ltd. 2017
J. Bhaumik et al. (eds.), *Communication, Devices, and Computing*, Lecture Notes in Electrical Engineering 470, https://doi.org/10.1007/978-981-10-8585-7_13

antenna, utilizing spur lines and strip loading to the patch, and 87% size reduction with dual-band operation has been reported in this article. Effects of rectangular slot on the patch and ground plane have been shown in [6]. In this paper, Chakraborty et al. have found dual-band operation of microstrip patch antenna with 53.73% of size reduction. Some research articles have shown the analysis of the effects of slot cutting on the patch, with equivalent circuits and theoretical studies [7–9].

In this article, a study on the effects of a rectangular slot in non-radiating edge of a rectangular microstrip patch antenna on compactness has been presented. The parametric studies have been shown in this paper, and the results have been discussed with equivalent circuit of slotted patch. It has been investigated that increasing the length of the slot causes to increase current path length, and as a result, the antenna resonates at lower frequency which leads to size reduction of the antenna. With the optimum dimension of slot, the antenna resonates at 2.8 GHz frequency and the compactness has been calculated as 78% with respect to conventional antenna.

2 Antenna Design

The configuration of the reference antenna (Antenna 1) resonating at frequency $f_r = 5.7\,\text{GHz}$ is shown in Fig. 1. The length ($L = 14$ mm) and width ($W = 18$ mm) of Antenna 1 have been calculated from conventional rectangular microstrip patch antenna design equations. The substrate (Arlon AD300A) with dielectric constant $\varepsilon_r = 3$, loss tangent $\tan \delta = 0.002$, and thickness $h = 1.524$ mm has been taken for this design. The dimensions of the ground plane have been taken as more than three times of the patch size as if it behaves as an infinite ground plane, and the antenna

Fig. 1 Reference antenna (Antenna 1) configuration

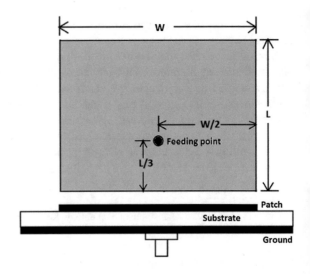

Fig. 2 Slotted antenna
(Antenna 2) configuration

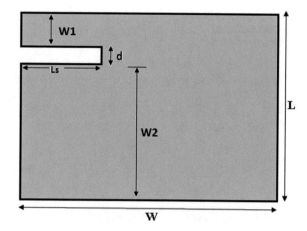

has been fed by a coaxial cable at optimum location to achieve the impedance very close to the characteristic impedance of the coaxial line, i.e., 50 Ω.

A slot has been incorporated on the patch in the non-radiating edge, and the antenna structure studied in this paper has been shown in Fig. 2, where d is the width of the slot and L_s is the length of the slot along the width of the antenna.

3 Equivalent Circuits

The equivalent circuit of the antenna will be the combination of rectangular patch and the notch-loaded patch. The notch dimension is $(L_s \times d)$.

A simple microstrip antenna is considered as a parallel combination of resistance (R_1), inductance (L_1), and capacitance (C_1). The equivalent circuit of the patch can be given as shown in Fig. 3

Fig. 3 Equivalent circuit of
reference antenna

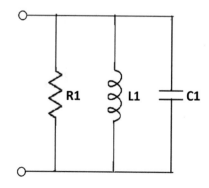

Fig. 4 Equivalent circuit
due to the effect of
notch-loaded patch

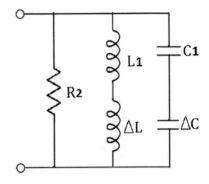

where R_1, L_1, C_1 can be defined as [10]

$$C_1 = \frac{\epsilon_0 \epsilon_{eff} L W}{2h} Cos^{-2}\left(\frac{\pi y_0}{L}\right), L_1 = \frac{1}{\omega^2 C_1}, R_1 = \frac{Q_r}{\omega c_1}$$

where L and W are the length and width of the rectangular patch antenna and h is
the thickness of the dielectric substrate of effective permittivity $\epsilon_{eff}.Q_r = \frac{C\sqrt{\epsilon_{reff}}}{4fh}$

$\epsilon_{reff} = \frac{\epsilon_r+1}{2} + \frac{\epsilon_r-1}{2}\left(1 + 12\frac{h}{W}\right)^{-\frac{1}{2}}$ where the symbols have their usual meanings.

When a notch is incorporated in a patch, two currents are flowing in the patch: One
is the normal current and the other is the current around the notch. Due to insertion
of the notch in a patch, discontinuities occur for magnetic field as well as electric
field both which can be modeled as an additional inductance ΔL and additional
capacitance ΔC with the existing inductance and capacitance of the equivalent circuit
of the patch [9] shown in Fig. 4.

The additional inductance ΔL can be written as [11, 12],

$$\Delta L = \frac{Z_1 + Z_2}{16\pi f\, Cos^{-2}\left(\frac{\pi y_0}{L}\right)} tan\left(\frac{\pi f L_s}{c}\right)$$

where

$$Z_1 = \frac{120\pi/\sqrt{\epsilon_{reff}}}{\left(\frac{W_1}{h}\right) + 1.393 + 0.667\, ln\left(\frac{W_1}{h} + 1.444\right)}$$

$$Z_2 = \frac{120\pi/\sqrt{\epsilon_{reff}}}{\left(\frac{W_2}{h}\right) + 1.393 + 0.667\, ln\left(\frac{W_2}{h} + 1.444\right)}$$

Table 1 Variation of antenna characteristics with L_s where $d = 1$ mm

Slot length Ls (mm)	f_r (GHz)	Return loss (dB)	B.W (GHz)	Gain (dBi)
5	4.602	14.54	0.14	3.9
6	4.602	13.75	0.84	3.9
7	4.595	27.82	0.147	3.9
8	4.581	25.61	0.133	−0.1
9	4.356	12.1	0.035	1.8
10	3.817	13.57	0.049	1
11	3.565	43.51	0.042	−3
12	3.333	41.72	0.042	−4.1
13	3.123	35.15	0.035	−6.1
14	2.943	32.28	0.028	−7

The capacitance ΔC is calculated as gap capacitance by [11]

$$\Delta C = 2(W_s)\frac{\epsilon_0}{\pi}\left[ln\left(2\frac{1+\sqrt{k'}}{1-\sqrt{k'}}\right) + \ln Coth\left(\frac{\pi d}{4h}\right) + 0.013C_f\frac{h}{d}\right]Cos^2\left(\frac{\pi y_0}{L}\right)$$

where $k' = \sqrt{1-k^2}$, $k^2 = \dfrac{1+\frac{W_1}{d}+\frac{W_2}{d}}{\left(1+\frac{W_1}{d}\right)\left(1+\frac{W_2}{d}\right)}$

C_f is the fringe capacitance [13] and is given by

$C_f = \frac{1}{2}\left[\frac{\sqrt{\epsilon_{reff}}}{cZ_0} - \epsilon_0\epsilon_r\frac{W}{h}\right]$, Z_0 is the characteristic impedance of the microstrip patch.

4 Results and Discussion

The simulated results of the slotted antenna with variation of L_s and d have been shown in Tables 1, 2, 3, and 4. The parametric studies have been done with variation of L_s from 5 to 14 mm and variation of d from 1 to 4 mm. Slot length (L_s) versus resonant frequency variation graph has been shown in Fig. 5, and slot length (L_s) vs gain graph has been shown in Fig. 6.

The current distribution of the slotted antenna for $L_s = 5$ mm, $d = 3$ mm at 4.644 GHz and $L_s = 14$ mm, $d = 3$ mm at 2.878 GHz have been shown in Figs. 7 and 8, and the respective radiation patterns have been shown in Figs. 9 and 10.

Table 2 Variation of antenna characteristics with L_s where $d = 2$ mm

Slot length Ls (mm)	f_r (GHz)	Return loss (dB)	B.W (GHz)	Gain (dBi)
5	4.63	40.47	0.147	4
6	4.616	43.91	0.138	4
7	4.581	34.53	0.133	3.8
8	4.433	40.22	0.091	2.1
9	4.146	35.25	0.056	0.1
10	3.873	36.21	0.056	−1.8
11	3.614	33.29	0.042	−3
12	3.375	25.05	0.035	−3.1
13	3.165	40.89	0.028	−6
14	2.969	27.11	0.028	−7

Table 3 Variation of antenna characteristics with L_s where $d = 3$ mm

Slot length Ls (mm)	f_r (GHz)	Return loss (dB)	B.W (GHz)	Gain (dBi)
5	4.644	36.3	0.14	4
6	4.616	40.36	0.133	3.9
7	4.511	33.19	0.106	2.5
8	4.274	34.53	0.077	2.8
9	4.006	31.17	0.056	−0.1
10	3.74	30.33	0.049	−1.9
11	3.494	27.76	0.043	−3
12	3.277	26.6	0.035	−4.3
13	3.067	25.78	0.035	−6
14	2.878	21.28	0.035	−7.8

Table 4 Variation of antenna characteristics with L_s where $d = 4$ mm

Slot length Ls (mm)	f_r (GHz)	Return loss (dB)	B.W (GHz)	Gain (dBi)
5	4.658	30.9	0.147	4
6	4.588	57.31	0.126	3.8
7	4.405	33.59	0.091	2.2
8	4.153	29.38	0.07	1.5
9	3.894	29.69	0.056	−0.1
10	3.642	32.6	0.049	−2
11	3.41	30.14	0.042	−3.8
12	3.193	38.81	0.035	−5
13	2.997	29.74	0.028	−6
14	2.808	24.62	0.035	−7.8

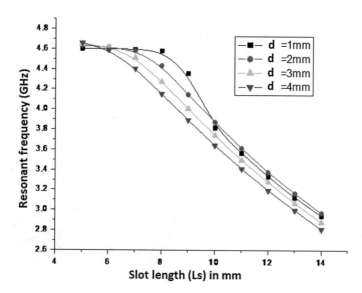

Fig. 5 Variation of resonant frequency with slot length

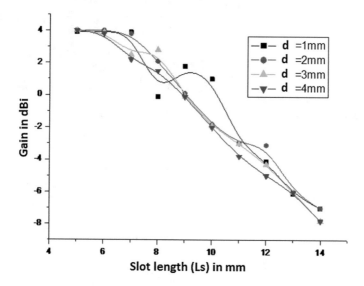

Fig. 6 Variation of gain with slot length

Fig. 7 Current distribution at 4.644 GHz

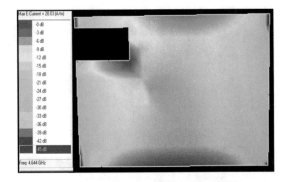

Fig. 8 Current distribution at 2.878 GHz

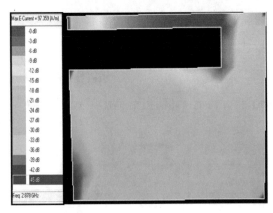

Fig. 9 Radiation pattern at 4.644 GHz

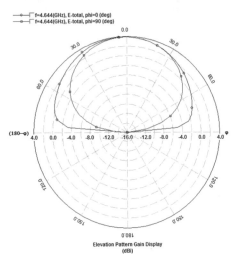

Fig. 10 Radiation pattern at 2.878 GHz

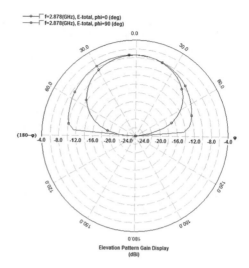

Elevation Pattern Gain Display
(dBi)

5 Conclusion

A study on the effects of rectangular slot on rectangular microstrip patch antenna has been performed and presented in this article. It has been discussed in the paper that frequency has been lowered considerably by increasing the length and width of the slot. The results have been discussed with approximated equivalent circuit of slotted patch, and some parametric studies have been done to show the variation of resonant frequency with slot length and width. With the optimum dimension of slot, almost 78% compactness has been found by simulation.

Acknowledgements The authors are grateful to the Department of Electronics and Communication Engineering, Central Institute of Technology, Kokrajhar, Assam, India, for giving the opportunity and financial support in carrying out this work.

References

1. K.L. Wong, *Planer Antennas for Wireless Communications* (Wiley, Hoboken, New Jersey, 2003)
2. S. Bhunia, D. Sarka, S. Biswas, P.P. Sarkar, B. Gupta, K. Yasumoto, Reduced size small dual and multi-frequency microstrip antenna. Microwave Opt. Technol. Lett. **50**, 961–965 (2008)
3. S. Bhunia, P.P. Sarkar, Reduced sized dual frequency microstrip antenna. Indian J. Phys. **83**, 1457–1461 (2009)
4. A. Roy, S. Bhunia, D. Chanda Sarkar, P.P. Sarkar, S.K. Chowdhury, Compact multi frequency strip loaded microstrip patch antenna with spur lines. Int. J. Microwave Wirel. Technol. **9**(5), 1111–1121 (2017)
5. A. Roy, S. Bhunia, D. Chanda Sarkar, P.P. Sarkar, Slot loaded compact microstrip patch antenna for dual band operation. Prog. Electromagn. Res. C (PIER C) **73**, 145–156 (2017)

6. U. Chakraborty, A. Kundu, S.K. Chowdhury, A.K. Bhattacharjee, Compact dual-band microstrip antenna for IEEE 802.11a WLAN application. IEEE Antennas Wirel. Propag. Lett. **13**, 407–410 (2014)
7. A. Mishra, P. Singh, N.P. Yadav, J.A. Ansari, B.R. Vishvakarma, Compact shorted microstrip patch antenna for dual band operation. Prog. Electromagn. Res. C **9**, 171–182 (2009)
8. Shivnarayan, S. Sharma, B.R. Vishvakarma, Analysis of slot-loaded rectangular microstrip patch antenna. Indian J. Radio Space Phys. **34**, 424–430 (2005)
9. Shivnarayan, B.R. Vishvakarma, Analysis of notch-loaded patch for dual-band operation. Indian J. Radio Space Phys. **35**, 435–442 (2006)
10. I.J. Bahal, P. Bhartia, Microstrip Antennas (Artech House, 1980)
11. Vijay K. Panday, Babau R. Vishvakama, Analysis of an E-shaped patch antenna. Microwave Opt Technol. Lett. **49**(1), 4–7 (2007)
12. C.A. Balanis, Advanced Engineering Electromagnetics (Wiley, 1989), pp. 449–452
13. J.A. Ansari, N.P. Yadav, P. Sing, A. Mishra, Compact half-U slot loaded shorted rectangular patch antenna for broadband operation. Prog. Electromagn. Res. M **9**, 215–226 (2009)

FPGA Implementation of OLS (32, 16) Code and OLS (36, 20) Code

Arghyadeep Sarkar, Jagannath Samanta, Amartya Barman
and Jaydeb Bhaumik

Abstract Orthogonal Latin square (OLS) codes are one type of one-step majority logic decodable (OS-MLD) error correcting code. These codes provide fast and simple decoding procedure. The OLS codes are used for correcting multiple cell upsets (MCU) which occur in semiconductor memories due to radiation-induced soft errors. OLS codes are derived from Latin squares and can be efficiently implemented on reconfigurable architectures like field programmable gate arrays (FPGA). This paper describes the construction of OLS codes from their parity check matrices and the method for increasing the data block size by extending the original OLS code. Here, double error correcting OLS (32, 16) code and OLS (36, 20) code have been designed and implemented on SRAM-based Xilinx FPGA. The synthesis results of area and delay of the encoder and decoder blocks are also presented. It is observed that extending the OLS codes will result in significant overhead in terms of the overall available resources and the delay of the codec circuits.

Keywords Error correcting code · Orthogonal Latin square · Memory
Soft error · FPGA

A. Sarkar
Jalpaiguri Government Engineering College, Jalpaiguri, India
e-mail: arghyadeep.sarkar@yahoo.com

J. Samanta (✉) · J. Bhaumik
Haldia Institute of Technology, Haldia, India
e-mail: jagannath19060@gmail.com

J. Bhaumik
e-mail: bhaumik.jaydeb@gmail.com

A. Barman
Wipro, Bangalore, India
e-mail: amartya.barman@gmail.com

© Springer Nature Singapore Pte Ltd. 2017
J. Bhaumik et al. (eds.), *Communication, Devices, and Computing*, Lecture Notes
in Electrical Engineering 470, https://doi.org/10.1007/978-981-10-8585-7_14

1 Introduction

Various expensive and sophisticated systems employed in avionics and spaces have been affected by serious threats like soft errors [1]. Although soft errors do not damage the hardware of the system, these can change the state of a datum or a signal which can lead to loss of information and can ultimately lead to functional failure [2]. Generally, single error correction–double error detection (SEC-DED) codes are thought to be sufficient to reduce the soft errors [3]. However, as technology scales down, radiation-induced soft errors can upset more than one memory cell [4]. Moreover, the radiation-induced errors can accumulate in a word corrupting it entirely. In such a scenario, the SEC-DED codes can be used in combination with bit interleaving [5]. Bit interleaving is a technique in which bits belonging to the same logical word are placed in different physical words, so that no two adjacent cells belong to the same physical word. Since bits belonging to the same logical word are physically apart, errors affecting multiple adjacent cells can easily be corrected by the SEC-DED codes. However, the above-mentioned technique has its fair share of limitations. First of all, the floor-planning and routing become more complex leading to greater overheads in area and power consumption [6]. Secondly, MCUs in registers and content addressable memories cannot be corrected by the aforementioned technique [7]. Thirdly, interleaving cannot protect memories suffering from MCUs caused due to several independent error events. A solution to these problems is using radiation-hardened cells, in which additional transistors are connected to the memory cells to make them robust and resilient to multiple upsets in the nodes of the cells [8–10].

An alternative to using such radiation-hardened cells is to use error correcting codes which can handle multiple errors [11, 12]. The advantage is offset in common multiple error correcting codes like Bose–Chaudhuri–Hocquenghem (BCH) codes and Reed–Solomon (RS) codes due to the high complexity of decoding algorithms resulting in higher overhead in delay and power [13]. This renders such codes to be less suitable to be used in high-speed memory elements.

One-step majority logic decodable (OS-MLD) codes are used to decide if an error has occurred in the bit by employing orthogonal parity checksums. OS-MLD decoders require lesser complexity and less delay as compared to BCH and RS decoders for the same data length. These codes are ideal for being implemented in high-speed circuits like caches and registers [14]. Although common OS-MLD codes, like difference set (DS) codes and Euclidean geometry (EG) codes, have been studied extensively for memory which are limited, these can provide a fixed error correcting capability for a given data size and can be implemented on limited data lengths.

A new class of OS-MLD codes, known as OLS codes [15], can be used to overcome the limitations of the DS and the EG codes. OLS codes use more number of parity bits as compared to other kinds of codes, but this drawback is mitigated by the fact that these codes are more versatile than the DS and the EG codes in terms of modularity and flexibility. These can also be implemented for protecting various data lengths and having a wide range of error correcting capabilities. Basic OLS codes can be used to protect data blocks of size m^2, where m is a positive integer. As the lengths of data blocks are in the powers of 2, so such unmodified OLS codes are useful to protect 16-bit and 64-bit data blocks. However, with slight modifications while keeping the code in line with its OS-MLD nature, the code can be extended to accommodate a wide range of data lengths, while using the same number of parity bits [16]. On the other hand, suitable measures can be taken to reduce the number of parity bits used for a particular extended data length [17]. Thus, the simplicity of the decoding process, low delay, modularity, flexibility and extensibility of the code outweighs the overhead due to the large number of parity bits. Hence, OLS codes can be successfully used to protect high-speed circuits like registers, caches and content addressable memories from soft errors.

In this paper, the properties and functionalities of orthogonal Latin square (OLS) codes are discussed. The basic OLS code is extended to accommodate larger data blocks keeping in line with the OS-MLD property. The paper also describes the method of construction of the parity check matrix or H-matrix of OLS codes and the way to design the encoder and the decoder from an H-matrix. Subsequently, the synthesis results of the VLSI implementation of the encoders and the decoders are presented using FPGA-based Virtex device family (Virtex-4 LX25, Virtex-5 LX50T and Virtex-6 LX75T).

In the post-introductory part, the properties and the construction of the H-matrix of OLS codes are summed up in Sect. 2, Sect. 3 deals with the gate-level design of the codes of the OLS codes, and the synthesis results of VLSI implementation of the codes are shown in Sect. 4. The paper concludes in Sect. 5 with a note on the future prospects and possibilities of this new class of multiple ECCs.

2 Orthogonal Latin Square Codes

OLS codes are a class of one-step majority logic decodable code with the fastest parallel decoding. In an OS-MLD code with t-error correcting capability, $(2t + 1)$ copies of each data bit are created from independent sources [18]. Out of the $(2t + 1)$ copies, 2t copies are derived from the 2t parity relations and one copy is received directly from the memory. So, each bit of a t-error correcting OLS code participates in 2t parity check equations, generating 2t parity check bits. These 2t parity check bits are fed into a majority logic circuit having 2t inputs. In the parity check bits, '1' indicates that an error has occurred and '0' indicates that no error has occurred. Now, for the t-error correcting OLS code, if the number of errors that occur for a particular data length during decoding of a particular bit is t, then it can be said that if an error corrupts the particular data bit, then the number of parity check bits the error can subsequently upset is $(t-1)$. Thus the remaining $(t+1)$ majority bits are not be affected which is shown in Fig. 1.

Hence, when an error occurs, the majority of the inputs and consequently the output of the majority logic decoder will be '1'. When no error occurs, the majority of the inputs and consequently the output of the majority logic decoder will be '0'. The output of the majority logic gate is XOR-ed with the received information bit to

Fig. 1 Simplified block diagram of OLS decoder using OS-MLD logic

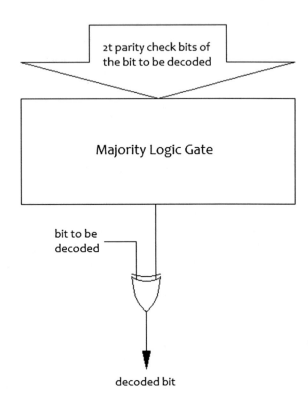

obtain the correct information bit. This decoding procedure is applied to all the bits of the data block to be corrected.

2.1 Construction of the Parity Check Matrix

OLS codes are derived from Latin squares. A Latin square of order m is a square array of size m × m with each row and column being a permutation of the numbers in the set {0, 1, 2, …, m − 1} in such a way that each number occurs only once in each row and column [19]. If the OLS code used for protecting data of k-bits length be derived from Latin squares of dimension m, then the relationship between k and m is $k = m^2$. The number of parity bits for the t-error correcting OLS code is 2tm. The general form of a parity check matrix of an OLS code from which the 2tm check bit equations originate is shown in Fig. 2 [15].

The M_1, M_2, M_3, …, M_{2t} are the sub-matrices of H having dimension m × m^2. The M_1 and M_2 have their usual forms as shown in Fig. 3. The number of 1's in each row of M_1 is equal to m and 'I_m' in M_2 are identity matrices of size m × m. The rest of the sub-matrices, i.e. M_3, M_4, …, M_{2t} are derived from Latin squares of order m

$$\mathbf{H} = \left[\begin{array}{c|c} \mathbf{M_1} & \\ \mathbf{M_2} & \\ \mathbf{M_3} & \mathbf{I_{2tm}} \\ \cdot & \\ \cdot & \\ \cdot & \\ \mathbf{M_{2t}} & \end{array}\right]$$

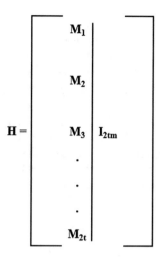

Fig. 2 Parity check matrix of a t-error correcting OLS code

$$\mathbf{M_1} = \begin{bmatrix} 1 & 1 & \dots & 1 & & & & & & & & & \\ & & & & 1 & 1 & \dots & 1 & & & & & \\ & & & & & & & & \dots & & & & \\ & & & & & & & & & 1 & 1 & \dots & 1 \end{bmatrix}$$

$$\mathbf{M_2} = \begin{bmatrix} \mathbf{I_m} & \mathbf{I_m} & \mathbf{I_m} & \dots & \mathbf{I_m} \end{bmatrix}_{\mathbf{m \times m^2}}$$

Fig. 3 First two sub-matrices M1 and M2 of the parity check matrix of an OLS code

that are pairwise orthogonal. Orthogonality of Latin square is the property by which if two Latin squares of the same order are superimposed, then every ordered pair formed by the elements of the squares is unique. Matrices S_1 and S_2 are two Latin squares that are pairwise orthogonal as they satisfy the aforementioned criteria [20].

$$S_1 = \begin{bmatrix} 0\ 1\ 2\ 3 \\ 1\ 0\ 3\ 2 \\ 2\ 3\ 0\ 1 \\ 3\ 2\ 1\ 0 \end{bmatrix} \tag{1}$$

$$S_2 = \begin{bmatrix} 0\ 1\ 2\ 3 \\ 2\ 3\ 0\ 1 \\ 3\ 2\ 1\ 0 \\ 1\ 0\ 3\ 2 \end{bmatrix} \tag{2}$$

For a t-error correcting OLS code, (2t − 2) orthogonal Latin squares are required for constructing the H-matrix. For a double error correcting code, t = 2. So, two pairwise orthogonal Latin squares are required in this case. Similarly, for a triple error correcting code, four pairwise orthogonal Latin squares are required. For a Latin square having m distinct elements, there exist m incidence matrices, each with respect to one element of the Latin square. The incidence matrices are of the same dimensions as the original Latin square, but in place of having m distinct elements or numbers, they are composed of 2 elements 0 and 1. To obtain the incidence matrix of a Latin square with respect to a particular value, the cells of the Latin square containing the value are substituted by '1' and the rest of the cells are substituted by '0'. The incidence matrix of the Latin square (1) with respect to the element '0' is shown in matrix 3.

$$
I = \begin{bmatrix} 1 & 0 & 0 & 0 \\ 0 & 1 & 0 & 0 \\ 0 & 0 & 1 & 0 \\ 0 & 0 & 0 & 1 \end{bmatrix} \tag{3}
$$

In a similar manner, three other incidence matrices can be derived from the Latin square with respect to the other 3 elements, viz. 1, 2 and 3. If these incidence matrices are to be expressed in one-dimensional vector form, then the dimension of each array will be $1 \times m^2$, where 'm' is the number of distinct elements in the Latin square [15]. The vector form of the incidence matrix with respect to 0 will be [1 0 0 0 0 1 0 0 0 0 1 0 0 0 0 1]. Similarly, three other vectors can be derived from the Latin square with respect to the three other numbers. The sub-matrix M_3 is constructed by these four vectors, and each vector forms a row of the sub-matrix. Likewise, sub-matrix M_4 can be constructed from another 4×4 Latin square orthogonal to the above Latin square. The parity check matrix H thus constructed for a double error correcting (t = 2) OLS code having Latin square block size 'm' equal to 4 and size of data block 'k' equal to 16 is shown in Fig. 4.

The property that makes OLS codes highly versatile is that they can be extended by extending the data bit block according to the norms of OS-MLD [16]. OLS codes follow one-step majority logic decoding and so, each column in the data block part of the H-matrix shares at most a single '1' at a particular position with every other column. Each column contains 2t bits in the H-matrix. So, the extension of the code can be done by adding columns (and rows if necessary) so that the number of bits in each column remains 2t. The H-matrix of the OLS (36, 20) code, derived by extending the OLS (32, 16) code keeping in line with the OS-MLD logic is depicted in Fig. 5. In this manner, H matrices can be generated for OLS codes of various data block sizes. Four columns are added as per OS-MLD logic.

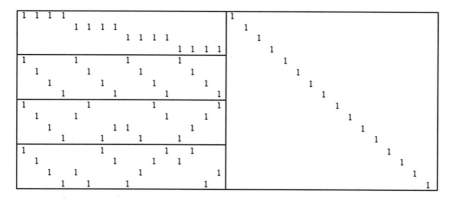

Fig. 4 Parity check matrix of OLS (32, 16) code with k = 16, m = 4 and t = 2

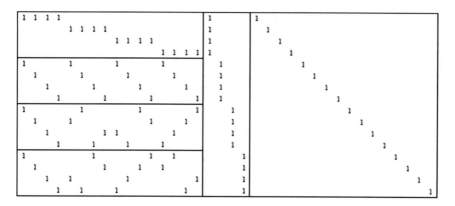

Fig. 5 Parity check matrix of extended OLS (36, 20) code with k = 20, m = 4 and t = 2

3   Design of OLS Codec

The encoder for an OLS code is very simple to design as the equations for the codeword calculation can be directly obtained from each row, and thus, the parity bits can be generated by passing the bits in each row through a modulo-2 adder. So, the number of parity bits is equal to the number of rows in the parity check matrix of the OLS code. Figure 6 shows the gate-level schematic diagram of the encoder for OLS (32, 16) code whose parity check matrix is shown in Fig. 4.

One-step majority logic decoding takes place in the decoder of an OLS code. For obtaining each bit for a codeword, the parity check equations in which the bit participates are recomputed in conjunction with the bit itself and a majority voting takes place among the parity check equations. If an error has occurred, the value of the resultant parity check equation or the syndrome bit becomes '1', else it remains '0'. If the majority of the syndrome bits become '1', then it is evident that an error has altered the state of the bit, and so, the bit needs to be corrected. The recomputing

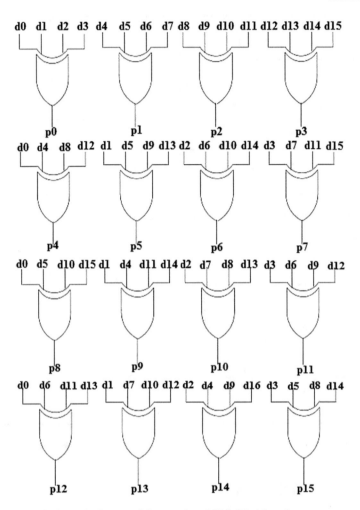

Fig. 6 Gate-level schematic diagram of the encoder of OLS (32, 16) code

of the parity check bits is done by modulo-2 adders or XOR gates having suitable number of inputs. The error can be corrected at the end of the majority logic circuit by using a 2-input XOR gate where one of the inputs is the bit itself and the other one is the output of the majority logic circuit. The detailed gate-level schematic diagram of the OLS (32, 16) decoder of the bit 'd$_0$' of a 16-bit data block (d$_0$ to d$_{15}$) is shown in Fig. 7. The 4-input majority logic circuit is encapsulated in a rectangular block. In the next section, two OLS codes have been implemented on FPGA platform.

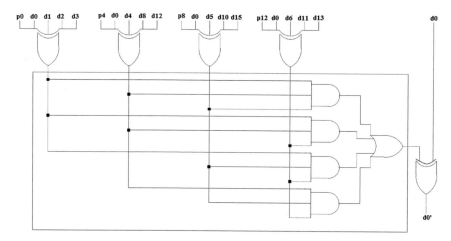

Fig. 7 Gate-level schematic diagram of OLS (32, 16) decoder for 16-bit data block

4 VLSI Implementation

All the functional blocks are represented using Verilog hardware description language. Both OLS (32, 16) and OLS (36, 20) are synthesized in FPGA-based Virtex-4, Virtex-5 and Virtex-6 device families.

Field programmable gate array (FPGA) is a solid-state device that can be hardware programmed to implement different functions by the end-user. FPGAs can be used for prototyping an idea into silicon in a very short time as they are completely fabricated and standard parts are tested by the manufacturer. The smallest unit for implementation of logic in an FPGA device is called a configurable logic block (CLB). CLBs contain a number of modules like look up tables (LUTs), flip-flops, multiplexers and gates. An LUT functions as a combinational circuit and stores the truth table of a Boolean function. The Virtex series of Xilinx are static RAM-based FPGA devices and thus are infinitely reconfigurable [21].

Two classes of OLS code, viz. OLS (32, 16) code and extended OLS (36, 20) code, are implemented in HDL on Virtex-4 LX25, Virtex-5 LX50T and Virtex-6 LX75T devices. The encoder and decoder of the two OLS codes are mapped onto the aforementioned devices, and results of resource utilization and time delay are recorded for each case.

From Table 1, it is evident that the resource utilization in terms of the number of LUTs used in both encoders and decoders is the greatest in case of Virtex-4 and is similar in case of Virtex-5 and Virtex-6 devices. This happens due to the fact that Virtex-4 devices use 4-input LUTs in their CLBs, but Virtex-5 and Virtex-6 devices use 6-input LUTs in their CLBs. This allows the LUTs of Virtex-5 and Virtex-6 devices to compute larger functions in their LUTs, thereby decreasing the overall resource utilization of the devices. The total time delay decreases from Virtex-4 to Virtex-6 devices in both encoders and decoders. From the tables, it can also be

Table 1 FPGA synthesis results of OLS (32, 16) and OLS (36, 20) codec

Device	Block	Area in terms of LUTs		Delay (ns)	
		OLS (32, 16) code	OLS (36, 20) code	OLS (32, 16) code	OLS (36, 20) code
Virtex-4 (LX25)	Encoder	16	32	4.986	5.33
Virtex-5 (LX50T)		16	16	3.885	4.042
Virtex-6 (LX75T)		16	16	0.989	1.117
Virtex-4 (LX25)	Decoder	65	72	6.376	6.567
Virtex-5 (LX50T)		32	37	4.789	4.877
Virtex-6 (LX75T)		32	37	1.744	1.778

observed that there is a slight overhead in terms of resource utilization and total time delay in case of the extended OLS (36, 20) code as compared to the OLS (32, 16) code. Thus, there is a major trade-off between OLS code for larger data blocks and the resource utilization and time delay of encoder and decoder circuits.

5   Conclusion

Orthogonal Latin square codes are highly efficient to correct multi-bit errors without suffering significant delay, and the encoders and the decoders can be implemented in relatively simple circuits as compared to other existing multi-bit error correcting codes. Based on this work, the encoder and decoder of OLS codes can be implemented for larger data blocks and adapted to variety of data lengths for their high flexibility and modularity. As for the future possibilities, efficient optimization techniques can be incorporated to reduce the number of parity check bits and the encoders and decoders can be implemented in a pipelined structure to further decrease the area and delay overheads while correcting errors in large data blocks.

References

1. R.C. Baumann, Radiation-induced soft errors in advanced semiconductor technologies. IEEE Trans. Device Mater. Reliab. **5**(3), 301–316 (2005)
2. B. Cooke, R. Muller, Error correcting codes. MIT Undergrad. J. Math. **1**, 21–26 (1999)
3. M.Y. Hsiao, A class of optimal minimum odd-weight-column SEC-DED codes. IBM J. Res. Dev. **14**, 395–401 (1970)

4. A. Dutta, N.A. Touba, Multiple bit upset tolerant memory using a selective cycle avoidance based SEC DED-DAEC code, in *Proceedings of IEEE VLSI Test Symposium* (2007), p. 349–354
5. S. Baeg, S. Wen, R. Wong, SRAM interleaving distance selection with a soft error failure model. IEEE Trans. Nucl. Sci. **56**(4), 2111–2118 (2009)
6. S. Baeg, S. Wen, R. Wong, SRAM interleaving distance selection with a soft error failure model. IEEE Trans. Nucl. Sci. **56**(4 (part 2)), 2111–2118 (2009)
7. S. Baeg, S. Wen, R. Wong, Minimizing soft errors in TCAM devices: a probabilistic approach to determining scrubbing intervals. IEEE Trans. Circuits Syst. I **57**(4), 814–822 (2010)
8. T. Calin, M. Nicoladis, R. Velazco, Upset hardened memory design for submicron CMOS technology. IEEE Trans. Nucl. Sci. **43**(6), 2874–2878 (1996)
9. S.M. Jahinuzzaman, D.J. Rennie, M. Sachdev, A soft error tolerant 10T SRAM bit-cell with differential read capability. IEEE Trans. Nucl. Sci. **56**(6), 3768–3773 (2009)
10. R. Rajaei, B. Asgari, M. Tabandeh, M. Fazeli, Design of robust SRAM cells against single event multiple effects for nanometer technologies. IEEE Trans. Device Mater. Reliab. **15**(3), 429–436 (2015)
11. P. Ankolekar, S. Rosner, R. Isaac, J. Bredow, Multi-bit error correction methods for latency-constrained flash memory systems. IEEE Trans. Device Mater. Reliab. **10**(1), 33–39 (2010)
12. G. Neuberger, D. Kastensmidt, R. Reis, An automatic technique for optimizing Reed-Solomon codes to improve fault tolerance in memories. IEEE Des. Test Comput. **22**(1), 50–58 (2005)
13. G.C. Cardarilli, M. Ottavi, S. Pontarelli, M. Re, A. Salsano, Fault tolerant solid state mass memory for space applications. IEEE Trans. Aerosp. Electron. Syst. **41**(4), 1353–1372 (2005)
14. S. Lin, D.J. Costello, *Error Control Coding*, 2nd edn. (Prentice-Hall, Englewood Cliffs, NJ, USA, 2004)
15. M.Y. Hsiao, D.C. Bossen, R.T. Chien, Orthogonal Latin square codes. IBM J. Res. Develop. **14**(4), 390–394 (1970)
16. P. Reviriego, S. Pontarelli, A. Sánchez-Macián, J.A. Maestro, A method to extend orthogonal Latin square codes. IEEE Trans. Very Large Scale Integr. Syst. **22**(7), 1635–1639 (2014)
17. P. Reviriego, S. Liu, A. Sánchez-Macián, L. Xiao, J. Maestro, A scheme to reduce the number of parity check bits in orthogonal Latin square codes. IEEE Trans. on Reliab. **66**(2)
18. M. Demirci, P. Reviriego, J. Maestro, Implementing double error correction orthogonal Latin squares codes in SRAM-based FPGAs. Microelectron. Reliab. **56**, 221–227 (2016)
19. J. Dénes, A.D. Keedwell, *Latin Squares and Their Applications* (Academic, San Francisco, CA, USA, 1974)
20. H.B. Mann, *Analysis and Design of Experiments* (Dover Publications, New York, 1949)
21. S. Hauck, A. De Hon Morgan, *Reconfigurable Computing: The Theory and Practice of FPGA-based Computation* (Kaufmann Publishers, Indian Reprint Edition, 2011)

Comparative Study of Wavelets for Image Compression with Embedded Zerotree Algorithm

Vivek Kumar and Govind Murmu

Abstract This paper presents study of different wavelets using the Embedded Zerotree Wavelet (EZW) algorithm, and their performance is analyzed for the application of image compression. The EZW is specially designed algorithm which uses zero tree property of wavelet transformed image to arrange the coefficients. These coefficients gives the progressively improved image information in order to pre-determined threshold. We have used Haar, Daubechies, Bi-orthogonal, Coiflet, and Symlets to perform discrete wavelet transform of a grayscale image. The effect of wavelet families has been analyzed on test images using measuring parameter: mean square error (MSE), peak signal-to-noise ratio (PSNR), maximum error, and compression ratio (CR). It is observed that using EZW algorithm, Coiflet and Symlet wavelet families produce uniform results in terms of MSE and PSNR.

Keywords Embedded Zerotree Wavelet (EZW) · Discrete wavelet transform (DWT) · Image compression · Mean square error (MSE) · Maximum error And peak signal-to-noise ratio (PSNR) · Compression ratio (CR)

1 Introduction

The growing demand of multimedia technologies requires larger bandwidth and space for transmission. They play a vital role as the use of image, audio, and video is common. It results in increase of use of data; hence, need of compressed information is preferred for different multimedia applications. Image compression [1] algorithms are widely used in most of the real-time transmissions and storage applications. Many image compression algorithms (lossless and lossy image compression) are available in the literature such as arithmetic coding, Huffman coding, predictive coding (lossless compression) and lossy predictive coding and transform domain coding (lossy compression). They are widely used for different image and video transmission per-

V. Kumar (✉) · G. Murmu
Department of Electronics Engineering, Indian Institute
of Technology (Indian School of Mines), Dhanbad, Jharkhand, India
e-mail: vivekismdhn@gmail.com

© Springer Nature Singapore Pte Ltd. 2017
J. Bhaumik et al. (eds.), *Communication, Devices, and Computing*, Lecture Notes in Electrical Engineering 470, https://doi.org/10.1007/978-981-10-8585-7_15

spectives which include Fourier transform, discrete cosine transform (DCT) [2], and discrete wavelet transform (DWT) [3]. Over the past few years, emergence of wavelet transform in image compression applications has shown substantial improvement in terms of picture quality at high compression ratio. It is mainly due to better energy compaction property of wavelet transform [3]. Many wavelet-based image compression algorithms have been developed using these properties of wavelet transform. They are progressive in nature such as Embedded Zerotree Wavelet algorithm (EZW) [4], Embedded Block Coding with Optimal Truncation (EBCOT) [5], Set Partitioning in Hierarchical Trees (SPIHT) [6], Adaptively Scanned Wavelet Difference Reduction (ASWDR) [7]. Although much of the research work [8, 9] had been published in the literature on image compression using EZW technique. This chapter will contribute discussion on performance of different wavelets particularly Haar, Daubechies, Bi-orthogonal, Coiflet, and Symlet with EZW algorithm with different test images.

The chapter is organized as follows: Wavelet transform in image compression is discussed in Sect. 2. Wavelets and their properties is discussed in Sect. 3. Short description on Embedded Zerotree Wavelet algorithm is discussed in Sect. 4. Experiment and results is presented in Sect. 5, conclusion in Sect. 6, and finally future work in Sect. 7.

2 Discrete Wavelet Transform in Image Compression

Wavelet transforms comprise of different wavelet families which have different set of scaling and wavelet functions. These wavelet families use different high-pass and low-pass filters for the analysis of different signals to perform image compression. With wavelet transform, image is first decomposed to an appropriate level using wavelets with high-pass and low-pass filters of that wavelet. As wavelets are small waves which are concerned with different patterns and properties, so they make different trade-offs between localization and smoothness, i.e., how compactly the basic functions are localized in space and how smooth they are. Some of the wavelet bases have fractal structure.

There are different types and families of wavelets whose properties may differ according to the following criteria [10]:

1. The support of wavelet and scaling functions or speed of convergence to zero in time domain, which defines the localization of scaling and wavelet function.
2. The symmetry which is useful in order to avoid image de-phasing.
3. The number of vanishing moments which is useful for compression.
4. The regularity, useful in smooth recovery of regular signals or images.
5. The existence of the scaling function ϕ.
6. The orthogonality or bi-orthogonality analysis.

Table 1 Properties of the selected wavelets

Properties	Haar	Daubechies	Symlets	Coiflets	Bi-orthogonal
Order (N)	1	2, 3, 4, ...	2, 3, 4, ...	1, 2, 3, ...	Nr, Nd (1, 2, 3, ...)
Orthogonal	Yes	Yes	Yes	Yes	No
Compact Support	Yes	Yes	Yes	Yes	Yes
Symmetrical	Yes	Yes	Yes	Yes	Yes
Support Width	1	$2N - 1$	$2N - 1$	$6N - 1$	$2Nr + 1, 2Nd + 1$
Filter Length	2	$2N$	$2N$	$6N$	Max (Nr, Nd) + 2
No. of vanishing moments	1	N	N	2N	Nr

Some important properties of wavelet functions in image compression applications are

1. Compact support which leads to efficient implementation.
2. Symmetry which is useful in avoiding de-phasing in image processing.
3. Regularity, degree of smoothness, and vanishing moments are related to filter order of filter length.

3 Wavelets and Their Properties

There are many wavelets available in the literature in which Haar wavelet, Daubechies wavelet (Db N), Coiflet wavelet (Coif N), Symlet wavelet (Sym N), and Bi-orthogonal wavelet (Bior Nr.Nd) are widely used for image compression. However, optimal wavelets are selected by choosing the appropriate properties which are required for the application. All of these wavelets are determined by filter order within the family except for Haar. Bi-orthogonal wavelets have filters with similar or dissimilar orders for decomposition (Nd) and reconstruction (Nr) within the same family. We have shown some common properties that each wavelet family has which is described in Table 1.

4 Image Compression Using EZW Algorithm

The EZW algorithm is specially designed compression algorithm by Shapiro's [4]. It encodes images in embedded fashion from their dyadic wavelet decomposition of the image. The goal of embedded coding is to generate a bit stream in such a fashion that allows encoder and decoder to terminate the process while retaining the best reconstruction quality at the available transmission rate.

The EZW algorithm is based on two important axioms:

1. As wavelet transform decomposes the image to a certain decomposition level which contains different sub-bands having different energy levels, these energy levels decrease as the scale decreases and the coefficients are smaller in higher sub-band than the lower. This property is utilized for the choice of progressive encoding for image compression.
2. The larger coefficients are more important than the smaller ones.

These two axioms are utilized by the Shapiro's algorithms by encoding the coefficients in several passes in decreasing order. After applying wavelet transform, an image can be represented using trees due to the subsampling that is performed in the transformation.

By choosing appropriate wavelet and level of decomposition, we apply wavelet transform and then from the decomposed coefficients initial threshold is selected. The encoding processes begin with initial threshold, and it is halved in further passes. The coefficients are scanned in the special order following raster pattern [10] and encoded by comparing the values with the threshold. If values are found to be greater and positive, they are coded as 'P', greater and negative codes as 'N', smaller but the corresponding child found to be greater than threshold coded as 'Z' and conditions for P, N, Z are not satisfied and then coded as 'T'. All coefficients are coded, dominant list is generated which is again arithmetically coded, and subordinate list is generated in form of '1' and '0'. Processes of dominant pass and subordinate pass are repeated by changing the threshold to its half of previous and replacing the scanned coefficients values by zero. These processes are repeated till target bits are achieved or all coefficients are coded. The information of dominant pass and subordinate pass is then passed to the transmission with initial threshold and different passes information. For the decoding process, the reverse process is followed.

5   Experiment and Results

The comparative analysis is carried out on three test images [11]: Lena, Cameraman, and Mandrill of size 256×256 using MATLAB. We have applied different available wavelets filters particularly Haar, Daubechies, Bi-orthogonal, Coiflets, Symlets, and Coiflet on the images to decompose at level 3. The transformed coefficients are then encoded and decoded using EZW compression technique. The performance of wavelets are analysed on reconstructed images using CR for lost amount of data, error is measured using MSE and maximum error, and image quality is measured using PSNR. The result analysis is shown in the Tables 2, 3, 4 and 5.

In Table 2, various results with selected wavelets are shown in which Db-5 wavelet produces maximum MSE and least PSNR for Cameraman image. Figure 1a–c shows reconstructed images with Db-5 wavelet.

Table 2 Results analyzed with Haar and Daubechies families

Image	Parameters	Wavelets						
		Haar	Db-2	Db-3	Db-5	Db-6	Db-7	Db-8
Lena image	MSE	154.8	139.6	131.1	134.4	248.3	139.5	252.2
	Max. Err.	102	95	88	78	123	85	122
	PSNR	26.23	26.88	26.96	26.52	24.18	26.68	24.11
	CR%	6.53	6.32	6.08	6.08	4.03	6.01	3.98
Cameraman image	MSE	128	139.7	135.6	308	145.8	146.4	309.8
	Max. Err.	113	109	105	160	102	105	161
	PSNR	27.03	26.88	26.81	23.23	26.49	26.48	23.22
	CR%	7.26	7.11	6.89	4.22	6.81	6.84	4.21
Mandrill image	MSE	270.5	248.9	240.7	238.2	238.1	237	234.2
	Max. Err.	83	90	82	99	95	82	84
	PSNR	23.81	24.17	24.32	24.36	24.36	24.38	24.43
	CR%	5.79	5.69	5.64	5.67	5.67	5.69	5.76

Table 3 Results analyzed with Coiflets family

Image	Parameters	Wavelets				
		Coif-1	Coif-2	Coif-3	Coif-4	Coif-5
Lena image	MSE	138.3	126.5	129.7	129.3	130.2
	Max. Err.	88	90	80	84	82
	PSNR	26.72	27.11	27	27.01	26.98
	CR%	6.3	6.02	5.95	5.97	5.99
Cameraman image	MSE	136	132.2	131.9	136.6	133.5
	Max. Err.	94	97	98	94	86
	PSNR	26.8	26.92	26.93	26.78	26.88
	CR%	7.01	6.81	6.84	6.72	6.82
Mandrill image	MSE	246.8	233.9	232.9	232.6	230
	Max. Err.	81	81	86	107	78
	PSNR	24.21	24.44	24.46	24.47	24.1
	CR%	5.71	5.68	5.74	5.7	5.67

Table 4 Results analyzed with Symlets family

Image	Parameters	Wavelets					
		SYM-2	SYM-3	SYM-4	SYM-5	SYM-6	SYM-8
Lena image	MSE	139.6	131.1	130.7	126.1	129.1	129.9
	Max. Err.	95	88	81	91	86	87
	PSNR	26.68	26.96	26.97	27.12	27.02	27
	CR%	6.33	6.09	6.15	6.01	6.07	6.03
Cameraman image	MSE	139.7	135.6	133.1	132.3	134.9	136.1
	Max. Err.	109	105	103	107	95	96
	PSNR	26.68	26.81	26.89	26.92	26.83	26.79
	CR%	7.11	6.89	6.79	6.78	6.71	6.72
Mandrill image	MSE	248.9	240	236.5	232.7	234.2	230.7
	Max. Err.	89	82	92	78	110	100
	PSNR	24.17	24.32	24.39	24.46	24.44	24.5
	CR%	5.7	5.64	5.71	5.66	5.66	5.73

Table 5 Results analyzed with Bi-orthogonal family

Image	Parameters	Wavelets						
		Bior 1.1	Bior 2.2	Bior 2.8	Bior 3.1	Bior 3.7	Bior 4.4	Bior 6.8
Lena image	MSE	154.8	118.2	195.1	262.5	172	133.3	125.9
	Max. Err.	102	87	118	134	109	87	88
	PSNR	26.23	27.41	25.16	23.94	25.77	26.88	27.13
	CR%	6.34	6.75	4.62	5.4	5.35	5.8	6.03
Cameraman image	MSE	128.9	259.3	239.3	305.7	212.3	133.8	129.7
	Max. Err.	113	132	133	144	146	99	96
	PSNR	27.03	23.99	24.37	23.28	24.86	26.87	27
	CR%	7.27	4.59	4.86	5.66	5.01	6.57	6.78
Mandrill image	MSE	270.5	212.8	201.1	402.6	184.9	238.8	226.9
	Max. Err.	83	87	81	101	73	82	86
	PSNR	23.81	24.85	25.1	22	25.46	24.35	24.57
	CR%	5.8	6.94	7.28	6.48	9.57	5.51	5.83

Fig. 1 **a** Using Db-5 wavelet, PSNR = 26.52 and CR = 6.08%; **b** using Db-5 wavelet, PSNR = 23.23 and CR = 4.22%; **c** using Db-5 wavelet, PSNR = 24.36 and CR = 5.67%

Fig. 2 **a** Using Coif-3 wavelet, PSNR = 27 and CR = 5.95%; **b** using Coif-3 wavelet, PSNR = 26.93 and CR = 6.84%; **c** using Coif-3 wavelet, PSNR = 27 and CR = 5.95%

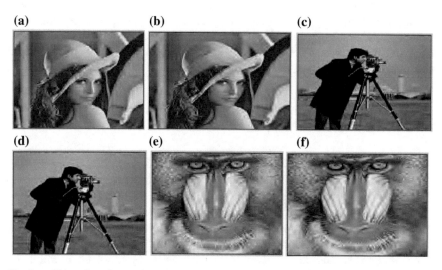

Fig. 3 **a** Using Sym-2 wavelet, PSNR = 26.68 and CR = 6.33%; **b** using Sym-5 wavelet, PSNR = 27.02 and CR = 6.02%; **c** using Sym-2 wavelet, PSNR = 26.68 and CR = 7.11; **d** using Sym-5 wavelet, PSNR = 26.92 and CR = 6.78%; **e** using Sym-2 wavelet, PSNR = 24.17 and CR = 5.33%; **f** using Sym-5 wavelet, PSNR = 24.44 and CR = 5.66%

(a)   **(b)**   **(c)**

(d)   **(e)**   **(f)**

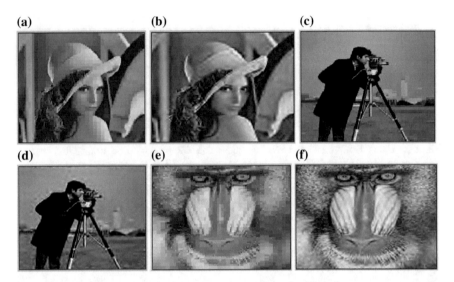

Fig. 4 **a** Using Bior 1.1 wavelet, PSNR = 26.23 and CR = 6.34%; **b** using Bior 3.7 wavelet, PSNR = 25.77 and CR = 5.35%; **c** using Bior 1.1 wavelet, PSNR = 27.03 and CR = 7.27%; **d** using Bior 3.7 wavelet, PSNR = 24.86 and CR = 5.01%; **e** using Bior 1.1 wavelet, PSNR = 23.81 and CR = 5.8%; **f** using Bior 3.7 wavelet, PSNR = 25.46 and CR = 9.57%

Table 3 shows results with various Coiflet wavelet families in which MSE and PSNR are uniform. Fig. 2a–c shows reconstructed images using Coiflet family.

Table 4 shows results with various Symlet wavelet families in which MSE and PSNR are uniform. Fig. 3a–f shows reconstructed images using Symlet family.

Table 5 shows results with various Bi-orthogonal wavelet families in which MSE and PSNR have more variations in all test images. Fig. 4a–f shows reconstructed images using Bi-orthogonal family.

6 Conclusion

Based on the experimental result shown in Tables 2, 3, 4, 5, it is found that the compression ratio for the set of Coiflet and Symlet wavelets produces nearly the uniform compression ratio. MSE and PSNR remain lowest and highest, respectively, for the set of Coiflet and Symlet wavelets, than other wavelet used. This confirms that Symlet and Coiflet produce better visual result than the other wavelet used with EZW algorithm. However, the MSE and PSNR with Bi-orthogonal wavelet have more variations because of difference in the analysis and synthesis filters within the same wavelet.

7 Future Work

For future work, more images can be considered specially the color images which have not considered in current discussion and different decomposition and thresholding techniques should be considered as different images have different properties, e.g., different edge characteristics that can be modelled with wavelets that may produce some important results. Other wavelets may be considered with EZW compression technique.

Acknowledgements Authors acknowledge the suggestions of reviewers to improve this paper. They also acknowledge IIT (ISM) Dhanbad for providing financial support to present this paper in esteemed conference.

References

1. R.C. Gonzalez, R.E. Woods, *Digital Image Processing*, 3rd edn. (2007)
2. S. Saha, Image compression—from DCT to wavelets: a review. Crossroads **6**(3), 12–21 (2000)
3. R.M. Rao, *Wavelet Transforms: Introduction to Theory and Applications*, (Pearson Education India, 1998)
4. J.M. Shapiro, Embedded image coding using zerotrees of wavelet coefficients. IEEE Trans. Signal Process. **41**(12), 3445–3462 (1993)
5. David Taubman, High performance scalable image compression with EBCOT. IEEE Trans. Image Process. **9**(7), 1158–1170 (2000)
6. Amir Said, William A. Pearlman, A new, fast, and efficient image codec based on set partitioning in hierarchical trees. IEEE Trans. Circuits Syst. Video Technol. **6**(3), 243–250 (1996)
7. J.S. Walker, T.Q. Nguyen, Adaptive scanning methods for wavelet difference reduction in lossy image compression, 2000, in *Proceedings 2000 International Conference on Image Processing*, vol. 3 (IEEE, 2000)
8. S.M. Jog, S.D. Lokhande, Embedded zero-tree wavelet (EZW) image CODEC, in *Proceedings of the International Conference on Advances in Computing, Communication and Control* (ACM, 2009)
9. A.P. Singh, B.P. Singh, A comparative study of improved embedded zerotree wavelet image coder for true and virtual images, in *2012 Students Conference on Engineering and Systems (SCES)* (IEEE, 2012)
10. M. Michel et al., (eds.), *Wavelets and Their Applications* (Wiley, 2013)
11. Standard test images (a set of images) found frequently in the literature. http://www. imageprocessingplace.com/downloads_V3/root_downloads/image_databases/standard_test_images.zip

Design of Compact Wideband Folded Substrate-Integrated Waveguide Band-Pass Filter for X-band Applications

Nitin Muchhal, Abhay Kumar, Arnab Chakrabarty and Shweta Srivastava

Abstract The paper proposes the design of X-band-pass filter. It has wide pass band with miniaturization of size using folded substrate-integrated waveguide (FSIW) technique. The paper also shows the effect of the slots in central conducting septum of FSIW on resonant frequency, impedance bandwidth, and return loss. In this paper, two different filters are simulated using HFSS v13 by introducing C and E slot in central metallic septum, respectively. The results so obtained are compared depicting the effect of introducing slot shapes on the performance of the band-pass filter. The result shows better performance (in terms of bandwidth, return loss, and compactness) of E slot FSIW as compared with C slot FSIW.

Keywords Folded substrate-integrated waveguide · Band-pass filter · C shape slot · E shape slot

1 Introduction

For past few decades, there has been ever-increasing demand and growth in military and commercial applications for microwave components having the attributes of compact size, low profile, large bandwidth, etc. For this reason, communication systems are expanding rapidly to higher frequency ranges and for past few years there has been rapid growth, tremendous demand, and immense prospects in microwave region (bands), especially C-band, X-band, and Ku-band. These bands have wide

N. Muchhal (✉) · A. Kumar · A. Chakrabarty · S. Srivastava
Department of E.C.E., Jaypee Institute of Information Technology, Noida, India
e-mail: nmuchhal@gmail.com

A. Kumar
e-mail: abhay.kumar@jiit.ac.in

A. Chakrabarty
e-mail: arnab.chk@hotmail.com

S. Srivastava
e-mail: shweta.srivastava@jiit.ac.in

© Springer Nature Singapore Pte Ltd. 2017
J. Bhaumik et al. (eds.), *Communication, Devices, and Computing*, Lecture Notes in Electrical Engineering 470, https://doi.org/10.1007/978-981-10-8585-7_16

applications [1] in marine, military application, satellite communication, weather monitoring, etc. The conventional rectangular waveguide (RWG) structures have advantages of its low loss and high Q. However, they are tedious to design, integrate with planar circuits, and expensive in mass production. Also higher frequencies prevent application of planar technology due to high transmission losses. The prospective solution for overcoming this problem at high frequency is substrate-integrated waveguide (SIW) technology [2].

The present paper proposes the design of miniaturized wide-band BPF with the help of folded SIW technology. The folded substrate-integrated waveguide (FSIW) takes the advantages of easy integration and compact size as compared to the conventional SIW structures. The proposed filter operates in X-band region (8–12 GHz) of microwave range. X-band is widely used for the military applications for beyond line-of-sight communication. For past few years, there has been increasing research on enhancing the bandwidth and miniaturization of SIW band-pass filters for microwave and millimeter-wave applications.

Hao et al. [3] proposed and designed a super-wideband band-pass filter by combining different types of periodic structures into the SIW. This was achieved by incorporating different kinds of periodic structure such as defect ground structures (DGSs) and electromagnetic bandgap structures (EBGs) with SIW. Alkhafaji et al. [4] proposed a compact band-pass filter with wide bandwidth by embedding two semi-circular slots in SIW geometry from the output and input sides, respectively. It had applications in X-band communication. Wong et al. [5] proposed a compact wideband filter by locating four viases in form of rotationally even design inside the SIW cavity. Danaeian et al. [6] proposed a band-pass filter by loading T-shape slots of various sizes on the upper conducting plane of the SIW structure and achieved wide bandwidth with reduction in size. The present paper focuses on improvement of bandwidth of folded SIW band-pass filter by introducing two different kinds of slot (C slot and E slot) on the central metal septum of the FSIW.

2   Design of Substrate-Integrated Waveguide (SIW)

Substrate-integrated waveguides (SIWs) are planar structures [7] which are fabricated by using two periodic rows of cylindrical vias implanted in a dielectric substrate that electrically unite both parallel conducting plates as shown in Fig. 1. SIW structure exhibits propagation characteristics like that of conventional waveguides and maintains most of the benefits of classical metallic waveguides [8]. Here the parameters P and D are the period of via holes and diameter of the metallic via, respectively. W_{SIW} is the distance between the rows of the centers of vias.

Fig. 1 Basic SIW structure

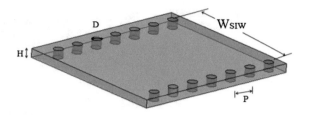

2.1 SIW Design Formulas

From [9], equivalent width of dielectric-filled rectangular waveguide

$$W_{EQ} = \frac{c}{2 f c \sqrt{\varepsilon_r}} \tag{1}$$

Width of SIW,

$$W_{SIW} = W_{EQ} + \frac{D^2}{0.95 P} \tag{2}$$

In addition for selecting P and D, following inequalities should be satisfied

$$P < 4D \text{ and } P \le \frac{\lambda_O}{2} \sqrt{\varepsilon_r} \tag{3}$$

From above design formulas, the design parameters for cutoff frequency $f_c = 12$ GHz, dielectric constant $\varepsilon_r = 2.55$ and height of SIW H = 1.6 mm, we get P = 1.2 mm, D = 0.80 mm, and $W_{SIW} = 8.30$ mm.

3 Design of Folded Substrate-Integrated Waveguide (FSIW)

SIW has several advantages, e.g., less leakage losses, better immunity to electro-magnetic interference, cheap, lightweight. However, compared with stripline or microstrip components, SIW has disadvantage of larger width for same circuits. In order to overcome this problem, the concept of folded substrate-integrated waveg-uides (FSIWs) was introduced by [10]. To miniaturize the SIW components, various techniques have been reviewed by [11]. Figure 2a, b show the top and side view of a (C type) folded SIW with transition. The transition is employed to implement the mode transformation between the stripline and the central metal septum of the FSIW. Here W_{FSIW} is the width of folded SIW which reduces to half of conventional SIW, L_{FSIW} is length of FSIW, G is the gap between the central metal septum and the right sidewall. H_{FSIW} is the total height of folded SIW and is twice height of conventional

Fig. 2 **a** Top view of FSIW, **b** side view of FSIW

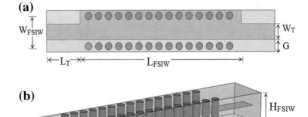

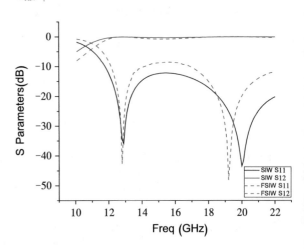

Fig. 3 Comparison of S-parameters of SIW and FSIW

SIW. Therefore, design parameters of FSIW for dielectric permittivity $\varepsilon_r = 2.55$ and cutoff frequency $f_c = 12$ GHz are $W_{FSIW} = 4.15$ mm, $H_{FSIW} = 3.2$ mm, $G = 1.5$ mm, $L_{FSIW} = 19$ mm, length of rectangular transition $L_T = 4$ mm, width of transition $W_T = 1.5$ mm. We have taken the end frequency point as cutoff as we are designing miniaturized band-pass filter for X-band applications. If a lower cutoff frequency is taken, then the operational frequency range of miniaturized band-pass filter may go below X-band. In addition, dielectric constant is taken as 2.55 as proposed filters are to be combined with an antenna to be used in transceiver section so it will be easier to design both elements with same substrate.

Figure 3 compares the return loss and insertion loss curves of SIW and FSIW using electromagnetic software HFSS [12]. From the figure, it is evident that FSIW has the same high-pass performance as SIW with same cutoff frequency.

Fig. 4 Folded SIW with C slot on central septum

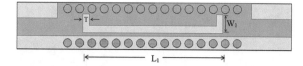

Fig. 5 S-parameter response of C slot FSIW

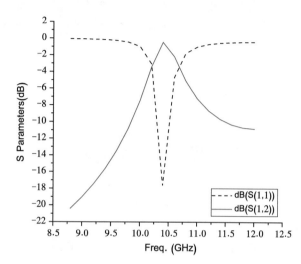

3.1 Design and Analysis of FSIW (with C Slot on Central Metal Septum of FSIW)

Since FSIW has the same high-pass performance as SIW, the band-pass function can be realized by introducing narrow slots on central metal septum of the FSIW. As shown in Fig. 4, a C slot is introduced on the central metal septum. Introducing a slot in central conductor will change the electrical length of the folded SIW which will change the response of FSIW from high pass to band pass. The geometry of modified FSIW is simulated using EM simulator HFSS v13 by taking dielectric substrate (Neltec NY9225) having dielectric constant 2.55 and height 3.2 mm. The loss tangent of the material is 0.0018. The parameters of the C slot have been optimized by parametric study of length and width of the slot and are given as follows: length of slot (L_1) is 14 mm, width of slot (W_1) is 2 mm, and thickness of slot (T) is taken as 0.65 mm.

Figure 5 shows the S-parameter response of C slotted FSIW. From the graph, it is evident that filter is resonating at 10.40 GHz with pass band's lower and upper frequencies as 10.30 GHz and 10.52 GHz, respectively, with a bandwidth of 0.22 GHz, −18.20 dB return loss, and −0.25 dB insertion loss. Figure 6 shows the current distribution of C slot FSIW.

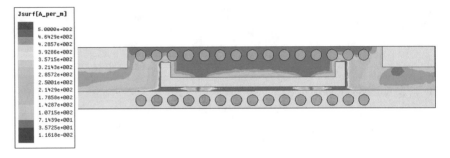

Fig. 6 Current distribution of C slot FSIW

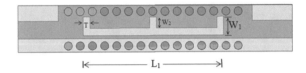

Fig. 7 Folded SIW with E slot on central septum

3.2 Design and Analysis of FSIW (with E Slot on Central Metal Septum of FSIW)

As shown in Fig. 7, an E slot is introduced on the central metal septum. The geometry is simulated using a dielectric substrate (Neltec NY9225) having dielectric constant 2.55, loss tangent of 0.0018, and height 3.2 mm. The dimensions of the filter are optimized using the EM simulator HFSS. E slot is made by introducing an additional small rectangular slot of width $W_2 = 1$ mm at the center of C slot as shown in Fig. 7. The thickness of each section of E slot remains same as T. Other dimensions L_1 and W_1 also remain the same as was taken for C slot FSIW.

Figure 8 shows the S-parameter response of E slotted FSIW. From the graph, it is evident that filter is resonating at 10 GHz with pass band's lower and upper frequencies as 9.75 GHz and 10.30 GHz, respectively, with a bandwidth of 0.65 GHz, -23 dB return loss, and -0.65 dB insertion loss. Figure 9 shows the current distribution of E slot FSIW. As evident from return loss graph, broader response is achieved in E slot FSIW due to addition of extra rectangular slot at middle section which introduces another resonance near the main resonance and thereby enhances the bandwidth.

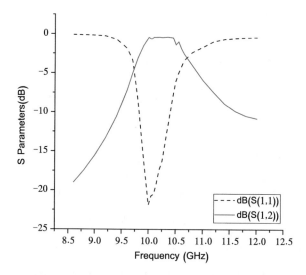

Fig. 8 S-parameter response of E slot FSIW

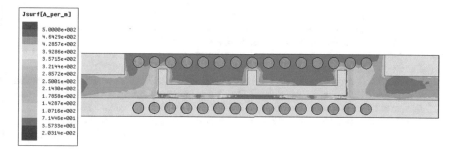

Fig. 9 Current distribution of E slot FSIW

4 Results and Conclusion

From the simulated S-parameter responses of both the filters, it is obvious that introducing E slot results in better performance of filter as compared to C slot in X-band application. The bandwidth of the proposed filter increases with the help of slots. The bandwidth is significantly increased by 66.15% in E slot FSIW. Also return loss is improved from −18.20 dB to −23 dB with degree of miniaturization by 16.67% for E slotted FSIW and 13.67% for C slotted FSIW band-pass filter, respectively. Further, the width of filter is reduced to half by virtue of folded topology.

References

1. D.M. Pozar, *Microwave Engineering*, 4th edn. (Wiley and Sons)
2. K. Wu, D. Deslandes, Y. Cassivi, The substrate integrated circuits—a new concept for high-frequency electronics and optoelectronics, in *6th International Conference on Telecommunications in Modern Satellite, Cable and Broadcasting Service, 2003 TELSIKS 2003*
3. Z.-C. Hao, W. Hong, J.-X. Chen, X.-P. Chen, W. Ke, Compact super-wide bandpass substrate integrated waveguide (SIW) filters. IEEE Trans. Microw. Theory Tech. **53**(9), 2968–2977 (2005)
4. A.N. Alkhafaji, A.J. Salim, J.K. Ali, Compact substrate integrated waveguide BPF for wideband communication applications, in *Progress in Electromagnetics Research Symposium Proceedings, Prague, Czech Republic, 6–9 July 2015*, pp. 135–140
5. S.W. Wong, R. Chen, K. Wang, Z.N. Chen, Compact SIW bandpass filter for wideband millimeter-wave applications, in *International Workshop on Microwave and Millimeter Wave Circuits and System Technology, Oct 2013, Chengdu, China*
6. M. Danaeian, A.G. Ashkezari, K. Afrooz1, A. Hakimi, A compact wide bandpass filter based on slotted substrate integrated waveguide (SIW) structure. J. Commun. Eng. **4**(2), 132–136 (2015)
7. U. Hiroshi, T. Takeshi, M. Fujii, Development of a laminated waveguide. IEEE Trans. Microw. Theory Tech. 2438–2443 (1998)
8. D. Deslandes, K. Wu, Single-substrate integration technique of planar circuits and waveguide filters. IEEE Trans. Microw. Theory Tech. **51**(2), 593–596 (2003)
9. Z. Kordiboroujeni, J. Bornemann, Designing the width of substrate integrated waveguide structures. IEEE Microwave Wirel. Compon. Lett. **23**(10), 518–522 (2003)
10. N. Grigoropoulos, B.S. Izquierdo, P.R. Young, Substrate integrated folded waveguides (SIFW) and filters. IEEE Microwave Wirel. Compon. Lett. **15**(12), 829–831 (2005)
11. N. Muchhal, S. Srivastava, Review of recent trends on miniaturization of substrate integrated waveguide (SIW) components, in *3rd IEEE International Conference on Computational Intelligence & Communication Technology (CICT), February 2017, Ghaziabad, India*
12. User Manual Ansys Inc., *High Frequency Structural Simulator (HFSS) Software, Version 13*

DCT-Based Gray Image Watermarking Scheme

Supriyo De, Jaydeb Bhaumik, Puja Dhar and Koushik Roy

Abstract In recent decades, information security is becoming a great challenge in the digital world. The authentication of digital image plays a vital role in secured digital data transfer. There exist several techniques for digital image watermarking to provide the security but standardization has not been done yet. In this paper, a DCT-based robust digital image watermarking scheme has been proposed. The scheme selects each block for DCT randomly. The scheme is verified by the different parameter such as peak signal-to-noise ratio (PSNR), normalized correlation (NC), structural similarity index (SSIM), and bit error rate (BER) for measuring image quality as well as embedding proficiency.

Keywords Image watermarking · DCT · PSNR · NC · SSIM · BER

1 Introduction

Watermarking is a technique to embed the intellectual property right to the digital multimedia for providing the identity of the source organization [1–3]. There are two different domains which are used for digital image watermarking: spatial domain and frequency domain. To implement a robust technique, researchers are liked to choose the frequency domain approach for digital image watermarking.

There are several tools used for the transformation [4, 5] such as discrete cosine transform (DCT), discrete wavelet transform (DWT), discrete Fourier transform

S. De (✉) · P. Dhar · K. Roy
Department of ECE, Saroj Mohan Institute of Technology, Guptipara, West Bengal, India
e-mail: supriyo.tech@gmail.com

P. Dhar
e-mail: puja.dhar.95@gmail.com

K. Roy
e-mail: roy.koushik94@gmail.com

J. Bhaumik
Department of ECE, Haldia Institute of Technology, Haldia, West Bengal, India
e-mail: bhaumik.jaydeb@gmail.com

© Springer Nature Singapore Pte Ltd. 2017
J. Bhaumik et al. (eds.), *Communication, Devices, and Computing*, Lecture Notes in Electrical Engineering 470, https://doi.org/10.1007/978-981-10-8585-7_17

(DFT), singular value decomposition (SVD). The basic concept of watermarking is to modify the frequency domain coefficient according to the bit pattern of message image.

A DCT-based approach was implemented to protect distribution right in image watermarking field in [6]. Several research groups used the combination of more than one transform to make the algorithm robust [7, 8]. In [9], 1-level DWT-based watermarking technique is being developed with the help of alpha blending technique. On the other hand, pixel-wise masking with DWT is being introduced in [10] which enhanced the image watermarking technique. SVD and Canny edge detection method both used for embedding the watermark logo in [11]. Agarwal et. al. [12] proposed a hybrid Fuzzy-BPN architecture for implementing a grayscale image watermarking technique which demands a good visual quality and robustness with respect to PSNR, SSIM, and NC.

In this paper, we have proposed a DCT-based image watermarking technique. Here, key-based non-repeating pseudorandom position generator is used to select the non-overlapping DCT block randomly. The scheme is tested with different parameters such as PSNR, SSIM, NC, and BER. A few common attacks also analyzed here with the said parameter, and the outcomes prove the robustness of the scheme.

The rest of the paper is organized as follows. In Sect. 2, proposed image watermarking scheme is described. Section 3 elaborately represents the experimental results. In Sect. 4, performance analysis of the proposed scheme is discussed and finally, conclusions are drawn in Sect. 5.

2 Proposed Scheme

The proposed image watermarking scheme is developed to hide the message image by using DCT tools. The entire scheme can be associated with three module: 1. pseudorandom non-repeated position generator, 2. DCT, and 3. embedding the message. The overall block diagram of the entire procedure is shown in Fig. 1.

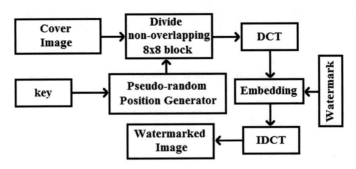

Fig. 1 Flowchart of embedding technique

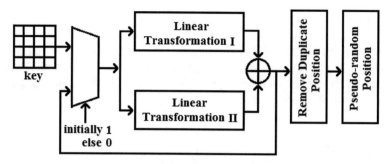

Fig. 2 Pseudorandom non-repeated position generator

2.1 Pseudorandom Non-repeated Position Generator

This block is responsible to generate the pseudorandom position for selecting the 8×8 block to store the bit information of the watermark image. It requires 128-bit private keys. There are two linear transformation blocks [13] which are used to permute and diffuse the bit pattern of keys to generate the pseudorandom position. Detailing operation is shown in Fig. 2.

2.1.1 Linear Transformation I (LT-1)

It is 128×128 binary tri-diagonal matrix. The upper and lower diagonals are ones, and the main diagonal is shown below.
0100100010001000010111110111101011001110000001110011110010111110
0101111001011000100000011100110101011101111101000001000100010010

2.1.2 Linear Transformation II (LT-2)

It is associated with the following structure of the matrix. T matrix is a (8×8) binary non-singular matrix and $T^L = I$, where I is a (8×8) identity matrix and $L = 255$. T matrix is as follows.

$$T_{all} = \begin{pmatrix} T^1 & T^2 & T^3 & .... & T^{15} & T^{16} \\ T^2 & T^4 & T^6 & & T^{30} & T^{32} \\ T^3 & T^6 & T^9 & & T^{45} & T^{48} \\ & ... & & & & \\ & ... & & & & \\ T^{15} & T^{30} & T^{45} & & T^{225} & T^{240} \\ T^{16} & T^{32} & T^{48} & & T^{240} & T^1 \end{pmatrix} \quad (a) \ and \ T = \begin{pmatrix} 1 & 1 & 0 & 0 & 0 & 0 & 0 & 0 \\ 1 & 1 & 1 & 0 & 0 & 0 & 0 & 0 \\ 0 & 1 & 0 & 1 & 0 & 0 & 0 & 0 \\ 0 & 0 & 1 & 1 & 1 & 0 & 0 & 0 \\ 0 & 0 & 0 & 1 & 0 & 1 & 0 & 0 \\ 0 & 0 & 0 & 0 & 1 & 1 & 1 & 0 \\ 0 & 0 & 0 & 0 & 0 & 1 & 0 & 1 \\ 0 & 0 & 0 & 0 & 0 & 0 & 1 & 1 \end{pmatrix} \quad (b) \quad (1)$$

2.1.3 Ex-OR

The outcome of *LT-1* and *LT-2* are now applied to the inequality detector by *Ex-OR* operation. The *Ex-ORed* result is then applied to the filter to remove the duplicate position and also the output fed as input of the *LT-1* and *LT-2* through multiplexer to get the next set of pseudorandom position.

2.2 Discrete Cosine Transform (DCT)

It is a well-known mathematical tool which may be used to transform an image from spatial domain to its frequency domain. It relates to the following mathematical equation for the image A.

$$B(p, q) = \alpha_p \, \alpha_q \sum_{m=0}^{M-1} \sum_{n=0}^{N-1} A(m, n) \cos\left(\frac{\pi(2m+1)p}{2M}\right) \cos\left(\frac{\pi(2n+1)q}{2N}\right) \quad (2)$$

where $0 \leq p \leq M - 1, 0 \leq q \leq N - 1$,

$$\alpha_p = \begin{cases} \frac{1}{\sqrt{M}}, & p = 0 \\ \sqrt{\frac{2}{M}}, & 1 \leq p \leq M - 1 \end{cases} \quad (a) \ and \ \alpha_q = \begin{cases} \frac{1}{\sqrt{N}}, & q = 0 \\ \sqrt{\frac{2}{N}}, & 1 \leq q \leq N - 1 \end{cases} \quad (b) \quad (3)$$

The inverse DCT can be done by the following equation for the frequency domain image B.

$$A(m, n) = \sum_{p=0}^{M-1} \sum_{q=0}^{N-1} \alpha_p \, \alpha_q \, B(p, q) \cos\left(\frac{\pi(2m+1)p}{2M}\right) \cos\left(\frac{\pi(2n+1)q}{2N}\right) \quad (4)$$

where $0 \leq m \leq M - 1$ and $0 \leq n \leq N - 1$

2.3 Embedding Watermark

A non-overlapping selected (using pseudorandom non-repeated position generator) 8×8 block of the cover image is used to hide the 1-bit information of the watermark image (binary image). A selected mid-frequency coefficient of DCT position is used here to hide the information. The total watermark information (bit pattern) is embedded in the same manner by selecting the different 8×8 block of the cover image. After hiding each bit, now the watermarked image is reconstructed by applying IDCT of each 8×8 block.

2.4 Extraction of Watermark

The validation or authentication of watermark is done using following steps.

1. Using the 128-bit private key generates the pseudorandom position.
2. Select the 8 × 8 blocks of the watermarked image.
3. Compute DCT for each 8 × 8 selected blocks.
4. Assign logic 1 to the watermark if the selected coefficient has positive polarity otherwise assign it by logic 0.
5. Reconstruct the watermark.

3 Experimental Results

Proposed image watermarking scheme is tested by several standard images. Watermark image size chosen for the test is 32 × 32. Figure 3a and 3b shows the cover image 'Lena' and watermark image (logo), respectively. The watermarked image and the extracted watermark are shown in Fig. 3c and Fig. 3d, respectively. The entire process has been done by using the following private key in both transmitting and receiving end.
Key = [01H 02H 03H 04H 05H 06H 07H 08H 09H 0BH 0BH 0CH 0DH 0EH 0FH 10H];

4 Performance Evaluation

The proposed scheme is tested with different properties of image watermarking such as imperceptibility, robustness, error probability. The scheme also checked by the said properties while a few possible attacks such as painting, cropping, and noise (salt and pepper) are considered.

(a)   **(b)**   **(c)**   **(d)**

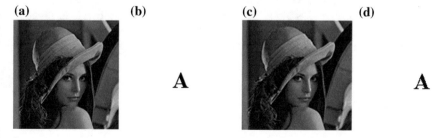

Fig. 3 **a** Cover image; **b** watermark image; **c** watermarked image; **d** extracted watermark

4.1 Imperceptibility

It indicates that how the embedded information makes changes to the host images. It is computed by different parameters. Here, we have considered the PSNR and SSIM to analyze the said property.

4.1.1 PSNR

Quality assessment of watermarked image with respect to its original cover image can be obtained by PSNR calculation [12]. The equation to compute PSNR is found in (5) and (6).

$$MSE = \frac{1}{J \times K} \left[\sum_{x=1}^{J} \sum_{y=1}^{K} \left(f(x, y) - f'(x, y) \right)^2 \right] \tag{5}$$

$$PSNR(dB) = 20 \log_{10} \frac{S_{\max}}{MSE} \tag{6}$$

where S_{max} is equal to 255 for 8-bit grayscale image; $J \times K$ is size of cover image; $f(x, y)$ is the original image (cover image) pixel value; $f'(x, y)$ is the watermarked image pixel value;

4.1.2 SSIM

It is a process which is used to measure the similarity in between cover image and watermarked image [12]. In (7), the mathematical detailing represented to calculate the SSIM.

$$SSIM = \frac{(2\mu_I \mu_{I'} + C_1)(2\sigma_{II'} + C_2)}{(\mu_I^2 + \mu_{I'}^2 + C_1)(\sigma_I^2 + \sigma_{I'}^2 + C_2)} \tag{7}$$

where μ_I and $\mu_{I'}$ are mean intensity of cover image and watermarked image; $C_1 = (J_1 L)^2$ and $C_2 = (J_2 L)^2$ are constants with L being the dynamic range of the grayscale image $(0 - 255)$ and $J_1 \ll 1$ and $J_2 \ll 1$ being small constants; $\sigma_I^2 = \frac{1}{n-1} \sum_{i=1}^{n} (I_i - \mu_I)^2$; $\sigma_{I'}^2 = \frac{1}{n-1} \sum_{i=1}^{n} (I_i' - \mu_{I'})^2$ and $\sigma_{II'}^2 = \frac{1}{n-1} \sum_{i=1}^{n} (I_i - \mu_I)(I_i' - \mu_{I'})$;

4.2 Robustness

This property can be verified by computing the NC in between extracted watermark and original watermark [12]. The computative expression is shown below.

Table 1 Comparative study—PSNR, SSIM, NC, BER

Image	Ref [6]				Ref [12]				Proposed scheme			
	PSNR	SSIM	NC	BER	PSNR	SSIM	NC	BER	PSNR	SSIM	NC	BER
Lena	46.9118	–	1	–	44.8919	0.9857	1	–	47.9349	0.9968	1	0
Barbara	46.8701	–	1	–	–	–	–	–	58.9749	1	1	0
Baboon	–	–	–	–	45.3276	0.9953	1	–	45.3007	0.9999	1	0
Peepers	–	–	–	–	44.2127	0.9865	1	–	55.2933	0.9999	1	0
Man	–	–	–	–	45.1247	0.9958	1	–	62.9057	1	0.9989	0.0009

$$NC(X, X^*) = \frac{\sum_{i=1}^{m} \sum_{j=1}^{n} [X(i, j) \cdot X^*(i, j)]}{\sum_{i=1}^{m} \sum_{j=1}^{n} [X(i, j)]^2} \tag{8}$$

where $m \times n$ is size of watermark image; $X(i, j)$ is the original watermark image; $X^*(i, j)$ is the extracted watermark image.

4.3 Error Probability

The reliability of watermarking scheme is tested by this parameter. The BER is computed in between the extracted watermark and the original watermark. It follows the standard equation in a communication system.

4.4 Comparative Study

A comparative study of all parameter is represented in Table 1 for few standard cover images. Numerical figures in Table 1 indicate the better performance of the proposed scheme. In this study, no attack is considered at transmitter, receiver, and network.

4.5 Attack and Analysis

Different types of well-known attacks, such as painting, cropping, salt and pepper noise, are considered at the watermarked image, and also, we have computed its corresponding parametric results. The attacked watermark images and extracted watermarks are shown in Fig. 4 for cover image 'Lena.' A complete picture of the parametric outcomes represents in Table 2. This study ensures the robustness of the proposed scheme.

Type of Attack   Watermark Image   Extracted Watermark

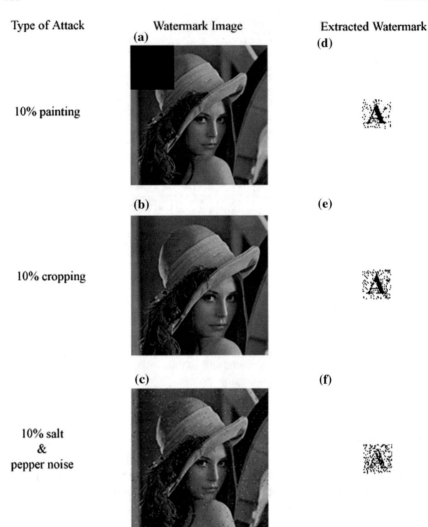

Fig. 4 **a** Watermarked image attacked by painting; **b** watermarked image attacked by cropping; **c** watermarked image attacked by salt and pepper noise; **d** extracted watermark from painting attack; **e** extracted watermark from cropping attack; **f** extracted watermark from salt and pepper noise attack

Table 2 Study of PSNR, SSIM, NC, BER for different kinds of attack

Image	10% painting attack				10% cropping attack				10% salt and peeper noise attack			
	PSNR	SSIM	NC	BER	PSNR	SSIM	NC	BER	PSNR	SSIM	NC	BER
Lena	18.9643	0.8956	0.8856	0.0986	14.5480	0.8228	0.8177	0.1592	25.3799	0.8014	0.7759	0.2256
Barbara	16.1200	0.8976	0.9060	0.0811	14.4235	0.8250	0.7950	0.1777	25.1553	0.8744	0.7848	0.2139
Man	18.8450	0.9022	0.9219	0.0674	16.0273	0.8320	0.8063	0.1670	24.9099	0.9529	0.7894	0.2119
Baboon	16.0468	0.8995	0.9049	0.0820	12.7022	0.8241	0.7928	0.1797	25.6078	0.9205	0.7894	0.2100
Peepers	15.7074	0.8979	0.9060	0.0811	12.5068	0.8235	0.7950	0.1768	25.1802	0.8306	0.7758	0.2285

5 Conclusions

The paper introduces a DCT-based new scheme for image watermarking. The 128-bit keys provide the additional security for authentication. The comparative study with respect to PSNR, SSIM, NC, and BER established the effectiveness of the proposed scheme. On the other hand, the outcome of different attacks also ensures the potentiality of the scheme as well.

References

1. I. Cox, J. Kilian, F.T. Leighton, T. Shamoon, Secure spreadspectrum watermarking for multimedia. IEEE Trans. Image Process. **6**, 1673–87 (1997)
2. R. Liu, T. Tan, A SVD-based watermarking scheme for protecting rightful ownership. IEEE Trans. Multimed. **4**(1), 121–128 (2002)
3. E.E. Abdallah, H.A. Ben, P. Bhattacharya, Improved image watermarking scheme using fast Hadamard and discrete wavelet transforms. J. Electron. Imaging **16**(3), 0330201–09 (2007)
4. S. Gupta, S. Jain, *A Robust Algorithm of Digital Image Watermarking Based on Discrete Wavelet Transform* (Department of Computer Science & Engineering, LNCT, Bhopal)
5. M.A. Mohamed, A.M. El-Mohandes, *Hybrid DCT-DWT Watermarking and IDEA Encryption of Internet Contents* (Electronics and Communication Engineering, Faculty of Engineering-Mansoura University, Mansoura, Dakhlia, Egypt)
6. G. Gupta, A.M. Joshi, K. Sharma, An efficient DCT based image watermarking scheme for protecting distribution rights, in *2015 Eighth International Conference on Contemporary Computing (IC3)* (IEEE, Noida, India), pp. 70–75
7. M. Malonia, S.K. Agarwal, Digital image watermarking using discrete wavelet transform and arithmetic progression technique, in *IEEE Students' Conference on Electrical, Electronics and Computer Science* (2016), p. 6
8. A. Akter, N.E. Tajnina, M.A. Ullah, Digital image watermarking based on DWT-DCT: evaluate for a new embedding algorithm, in *3rd International Conference on Informatics, Electronics & Vision* (2014), p. 6
9. A.P. Singh, A. Mishra, Wavelet based watermarking on digital image. Indian J. Comput. Sci. Eng. **1**(2), 86–91 (2011)
10. M. Barni, F. Bartolini, A. Piva, Improved wavelet-based watermarking through pixelwise masking. IEEE Trans. Image Process. **10**(2), 783–791 (2002)
11. A. Rajani, T. Ramashri, Image watermarking algorithm using DCT, SVD and edge detection technique. Int. J. Eng. Res. Appl. **1**(4), 1828–1834. ISSN 2248-9622
12. C. Agarwal, A. Mishra, A. Sharma, A novel gray-scale image watermarking using hybrid fuzzy-BPN architecture. Egypt. Inform. J. (Elsevier) 83–102 (2015). https://doi.org/10.1016/j.eij.2015.01.002
13. J. Bhaumik, S. De, A symmetric key based image encryption scheme. in *International Conference on Computing and Communication Systems* (North Eastern Hill University, Shillong, 2016), p. 8 (In press for published in Springer Lecture Notes in Networks and Systems)

Five-Input Majority Gate Design with Single Electron Nano-Device

Arpita Ghosh and Subir Kumar Sarkar

Abstract A five-input majority gate design using the single electron nano-device is presented in this work for the future ultra-dense integration. The majority gates are important part of the decision-making circuits based on voting. Being one of the nano-electronics devices single electron devices reduces the size of the complete circuit. The function of the five-input majority gate is tested by simulating the proposed circuit in SIMON simulator and analyzing the output waveforms.

Keywords Nano-electronic circuits · Single electron tunneling · Five-input majority gate · SIMON

1 Introduction

The nano-electronic devices [1, 2] facilitate a bunch of advantages in the form of size reduction, high speed, reduction in power requirement, higher packing density, and integration in the ultra-scale. Thus, the circuit implementation in the digital as well as in the analog domain has become very lucrative using different nano-electronic devices. Among the different nano-devices, single electron devices (SEDs) [3, 4] are one of the fascinating devices with unique Coulomb oscillation and Coulomb blockade [5, 6] characteristics. The Boolean logics are encoded using single electron devices with the presence and absence of the electrons (i.e., logic high and logic low). Another way of using the SED logic is by using the linear threshold gate [7–10] and weighted sum calculation. The second method reduces the number of tunnel junction requirement than the prior one thus reduces the power consumption of the circuit further. Several linear threshold logic-based circuit implementations [8, 9] with SEDs are reported in the past.

A. Ghosh (✉)
RCC Institute of Information Technology, Kolkata 700015, India
e-mail: arpita161@gmail.com

S. K. Sarkar
Jadavpur University, Kolkata 700032, India

© Springer Nature Singapore Pte Ltd. 2017
J. Bhaumik et al. (eds.), *Communication, Devices, and Computing*, Lecture Notes
in Electrical Engineering 470, https://doi.org/10.1007/978-981-10-8585-7_18

A majority logic gate [11] is mainly required in the voting-based decision circuits where the number of votes or same inputs is judged. It works according to the majority logic of number of inputs. It mainly consists of odd number of inputs. So that the decision based on the majority function can be easily chosen. For a n input (n should be odd number) majority gate if at least (n + 1)/2 inputs are high then the gate gives output as logic high otherwise the output remains at logic low. This work is mainly based on the design and implementation of a five-input-based majority gate with single electron nano-device. The circuit implementation and simulation are done using Monte Carlo-based simulator named SIMON [12] for simulating single electron nano-devices. Justification in support of the simulated results is also provided in the paper.

The paper is organized in the following way—first section deals with the overall introduction of the work and the single electron devices, while the second section elaborates the five-input majority gate implementation. The simulated results are shown in section three, whereas the fourth section concludes the complete work.

1.1   Single Electron Device

The tunnel junction, formed out of an extremely thin insulating material sandwiched between two metallic parts, is the main element of the single electron device. The single electron devices may be in the form of single electron box (SEB) or single electron tunneling junction, multiple tunnel junctions, or even a single electron transistor (SET). The working principle behind all the SEDs is same, i.e., controlled tunneling of number of electrons through the tunneling junction by the voltage across the tunnel junction. If the voltage across the junction crosses the threshold value, then the current flows due to tunneling of electrons. This specific voltage is here called the critical voltage. The range of the voltage for which the output current remains zero at 0 K temperature is known as the Coulomb blockade voltage. The circuit behaves as a stable one as no tunneling occurs during this above mentioned voltage range.

This Coulomb blockade and non-Coulomb blockade region concept or the switching of states from Coulomb blockade to non-Coulomb blockade and vice versa is mainly used for logical operation representations with single electron devices. Figure 1 depicts the drain current characteristics of the SET. The drain current (I_{DS}) is plotted against the drain to source voltage (V_{DS}) with 0 gate to source voltage (V_{GS}). The range of V_{DS} for which the drain current is 0 is marked in Fig. 1 as Coulomb blockade.

2   Five-Input Majority Gate Implementation

The majority gate produces output by considering all the input value conditions. If majority numbers of the inputs are at logic high level, then the output logic level is

Fig. 1 Single electron
device drain current with
drain voltage

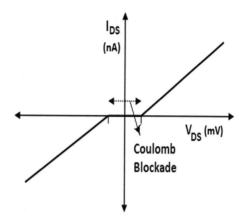

Fig. 2 Single electron
nano-device-based five-input
majority gate

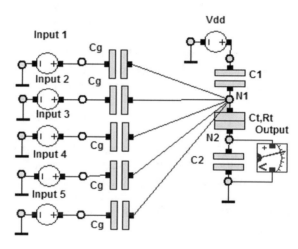

set at logic high otherwise low. The design of five-input majority gate using SED is shown in Fig. 2. The circuit design and simulation are done using SIMON simulator. Here one single electron tunnel junction is involved with junction resistance value of Rt and junction capacitance value of Ct. For the design, one bias capacitor C1 is connected with the bias voltage. The output or the load capacitor is denoted with C2. All the input signals are fed to node N1 through their corresponding input capacitors denoted with Cg. In this proposed design, five inputs are considered such as Input 1, Input 2, Input 3, Input 4, and Input 5. The output is taken from node N2. According to the majority gate with five inputs, the expression of the output is

$$Output = \overline{Input\ 1}(Output\_Input\ 1\_low) + Input\ 1(Output\_Input\ 1\_high)$$

$$(1)$$

where the overall output observed for two conditions of Input 1 Output_Input 1_low and Output_Input 1_high. The expressions for the Output_Input 1_low and Output_Input 1_high are shown in Eqs. (2) and (3)

$$Output\_Input\ 1\_low = \begin{pmatrix} \overline{Input\ 2}.\,Input\ 3.\,Input\ 4.\,Input\ 5+ \\ Input\ 2.\,\overline{Input\ 3}.\,Input\ 4.\,Input\ 5+ \\ Input\ 2.\,Input\ 3.\,\overline{Input\ 4}.\,Input\ 5+ \\ Input\ 2.\,Input\ 3.\,Input\ 4.\,\overline{Input\ 5}+ \\ Input\ 2.\,Input\ 3.\,Input\ 4.\,Input\ 5 \end{pmatrix} \qquad (2)$$

$$Output\_Input\ 1\_high = \begin{pmatrix} \overline{Input\ 2}.\,\overline{Input\ 3}.\,Input\ 4.\,Input\ 5 + \overline{Input\ 2}.\,Input\ 3.\,\overline{Input\ 4}.\,Input\ 5+ \\ \overline{Input\ 2}.\,Input\ 3.\,Input\ 4.\,\overline{Input\ 5} + \overline{Input\ 2}.\,Input\ 3.\,Input\ 4.\,Input\ 5+ \\ Input\ 2.\,\overline{Input\ 3}.\,\overline{Input\ 4}.\,Input\ 5 + Input\ 2.\,\overline{Input\ 3}.\,Input\ 4.\,\overline{Input\ 5}+ \\ Input\ 2.\,\overline{Input\ 3}.\,Input\ 4.\,Input\ 5 + Input\ 2.\,Input\ 3.\,\overline{Input\ 4}.\,\overline{Input\ 5}+ \\ Input\ 2.\,Input\ 3.\,\overline{Input\ 4}.\,Input\ 5 + Input\ 2.\,Input\ 3.\,Input\ 4.\,\overline{Input\ 5}+ \\ Input\ 2.\,Input\ 3.\,Input\ 4.\,Input\ 5 \end{pmatrix} \qquad (3)$$

The parameter values are Cg = 0.5 × Cj, Ct = 0.1 × Cj, C2 = 10 × Cj, and C1 = 10 × Cj. The tunneling junction resistance is chosen as 0.1 MΩ as it should be much greater than the quantum resistance (R_q).

$$R_t\ >\ R_q\ \approx\ 26\,k\Omega \qquad (4)$$

where $R_q = h/e^2$, h is Planck's constant and e is the single electronic charge. The power supply voltage is 0.1e/Cj, and the value of Cj is 1 aF. The size of the circuit depends on the area defined by the number of elements required for the circuit design. For this proposed circuit, total eight elements are required out of which seven are capacitors. The expression of the output function according to threshold logic output function can be written as

$$Output = sgn(Input\ 1 + Input\ 2 + Input\ 3 + Input\ 4 + Input\ 5 - 2.5) \qquad (5)$$

Equation (5) defines the relation of input and output using linear threshold logic function [10] where 2.5 is the threshold value calculated according to the threshold logic expressions [10].

3   Simulation Results of Five-Input Majority Gate

Two conditions of the Input 1 are considered—one is logic low (0 V) and another is logic high (16 mV). The other input signals are set according to Fig. 3a–d where they are plotted with respect to time.

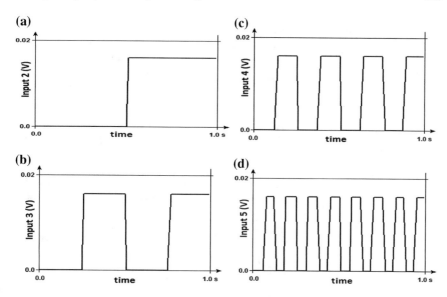

Fig. 3 Input signal waveforms for **a** Input 2, **b** Input 3, **c** Input 4, and **d** Input 5

The output signals are shown in Fig. 4a, b. For the two cases of the Input signal 1, two respective outputs are plotted. The first one (Fig. 4a) is simulated when the Input 1 is at logic high and, the second one (Fig. 4b) is for Input 1 with logic low. From Fig. 4a, it is clear that when Input 1 is high and Input 4, 5 are also high then total three out of five are high accordingly output is high. In the similar way in presence of high Input 1 when Input 3, 5 or 3, 4 or 3, 4, 5 or 2, 5 or 2, 4 or 2, 4, 5 or 2, 3 or 2, 3, 5 or 2, 3, 4 or 2, 3, 4, 5 are high same condition is satisfied and according to majority output also stays in logic high for all the mentioned combinations of input signals. Figure 4b depicts the output waveform for Input 1 at logic low. According to the figure, when Input 3, 4, 5 or 2, 4, 5 or 2, 3, 5 or 2, 3, 4 or 2, 3, 4, 5 are high but Input 1 is low as majority number of inputs are in logic high the output is also high. This clearly satisfies Eq. (2). The simulated output waveforms prove that logically the circuit is working as a five-input majority gate.

4 Conclusion

The successful design and implementation of the single electron nano-device-based five-input majority logic gate using SIMON simulator are furnished in this work. From the simulated results, it can be observed that with at least three or more than three numbers of input signals in the logic high level, the output of the designed circuit is also logic high. The output for all the possible input signal conditions is

Fig. 4 Output waveforms
for **a** Input 1 = logic high
and **b** Input 1 = logic low

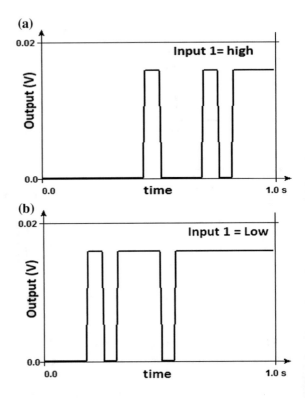

checked. The simulated input and output waveforms affirm that the proposed circuit
result is in close agreement with the desired one.

References

1. G. Frazier, *An Ideology for Nanoelectronics, in Concurrent Computations: Algorithms, Archi-tecture, and Technology* (Plenum Press, New York, 1988)
2. R.T. Bate et al., An overview of nanoelectronics. Texas Instrum. Technol. J. 13–20, July–Aug (1989)
3. K. Likharev, Single-electron devices and their applications. Proc. IEEE **87**, 606–632, Apr (1999)
4. Y. Takahashi, et. al., Silicon single-electron devices. J. Phys. Condens. Matter, R995–R1033 (2002)
5. T.A. Fulton, P.L Gammel, L.N. Dunkleberger, Determination of Coulomb-blockade resistances and observation of the tunneling of single electrons in small-tunnel-junction circuit. Phys. Rev. Lett. **67**, 3148–3151, Nov (1991)
6. D.V. Averin, K. Likharev, Coulomb blockade of tunneling, and coherent oscillations in small tunnel junctions. J. Low Temp. Phys. **62**, 345–372 (1986)
7. A. Ghosh, A. Jain, S.K. Sarkar, *Design and Simulation of Single Electron Threshold Logic Gate Based Programmable Logic Array*, vol. 10 (Procedia Technology, Elsevier, 2013), pp. 866–874

8. A. Ghosh, S.K. Sarkar, Comparative study of single electron threshold logic based and SET-CMOS hybrid based 1 bit comparator, in *Computational Science and Engineering* (Taylor & Francis Group, Dec 2016), pp. 235–238

9. J. Fernandez Ramos, J.A. Hidalgo Lopez, M.J. Martin, J.C. Tejero, A. Gago, A Threshold logic gate based on clocked couple inverters. Int. J. Electron. **84**(4), 371–382 (1998)

10. C.R. Lageweg, S. Cotofana, S. Vassiliadis, A linear threshold gate implementation in single electron technology, in *Proceedings IEEE Computer Society Workshop VLSI—Emerging Technologies for VLSI Systems* (Orlando, FL, Apr 2001), pp. 93–98

11. T. Oya, T. Asai, T. Fukui, Y. Amemiya, A majority logic device using an irreversible single-electron box. IEEE Trans. Nanotechnol. **2**(2), 15–22, Mar (2003)

12. C. Wasshuber, SIMON—A simulator for single-electron tunnel devices and circuits. IEEE Trans. Comput. Aided Design Integr. Circuits Syst. **16**(9), 937–944 (1997)

Post-layout Power Supply Noise Suppression and Performance Analysis of Multi-core Processor Using 90 nm Process Technology

Partha Mitra and Jaydeb Bhaumik

Abstract This article deals with accurate budget and placement of decoupling capacitance (decap) based on the supply noise. Decaps are placed near the noisy modules and effectively reduce the supply drop. The novelty of this paper lies in exhaustive estimation of Ldi/dt drop and IR drop for the complete circuit, followed by an algorithmic estimation and appropriate allocation of decaps with an effort to keep power, delay, and noise performance to its best. In this work, the suppression in supply noise, power consumption. and propagation delay parameters with decap allocation for 512-point FFT core are investigated and satisfactory results are obtained.

Keywords Decoupling capacitor (decap) · FFT processor · Power supply noise (PSN) · Power distribution network (PDN) · Application-specific integrated circuit (ASIC)

1 Introduction

Power distribution network in VLSI domain is of design and analysis of electrical network to the on-chip components of on-chip conductors. With the increase in functional modules, the packing density in a chip increases proportionally. Thus, supply noise varies almost proportionally with the increase of the processing cores or packing density of the chip circuitry.

Research papers [1–6] shows the analysis of supply noise and optimization of power distribution network (PDN). Decap is a commonly used technique for sup-

P. Mitra (✉)
Department of Electronics and Communication Engineering, Brainware
Group of Institutions-S.D.E.T., Barasat, Kolkata 700125, India
e-mail: partha_mitra_kgp@yahoo.co.in

J. Bhaumik
Department of Electronics and Communication Engineering, Haldia Institute
of Technology, Haldia 721657, India
e-mail: bhaumik.jaydeb@gmail.com

© Springer Nature Singapore Pte Ltd. 2017
J. Bhaumik et al. (eds.), *Communication, Devices, and Computing*, Lecture Notes
in Electrical Engineering 470, https://doi.org/10.1007/978-981-10-8585-7_19

pression of power supply noise. Generally, decap placement is done wherever white space (WS) is available inside the chip. We investigate decap deployment through module-wise analysis as well as suppression of supply noise for multi-core architectures. Major contributions in this paper are:

(i) Suppression of supply noise, power consumption, and propagation delay with and without decap.
(ii) 512-point FFT processor is analyzed as test circuit using CAD tools.

Rest of the paper is as follows. Introduction of power distribution network (PDN) and supply noise is discussed in Sect. 2. In Sect. 3, budget and placement of decaps have been discussed. In Sect. 4, a CAD methodology of our work for FFT processor is discussed. In Sect. 5, implementation and analysis of the results are elaborated. Finally, this paper is concluded in Sect. 6.

2   Power Distribution Network and Supply Noise

Power distribution network contains inductive, resistive, and capacitive components. In modern era, the packing density in a single chip has increased significantly. This has significant effect on design parameters and the supply noise. The variation of the supply voltage from the desired voltage level is known as the power supply noise. Sources of supply noise are discussed below.

2.1   Ldi(t)/dt Noise

In case of the on-chip power supply, the Ldi/dt noise is usually attributed to the package inductance. Alternatively, there have been some recent differing opinions on whether the on-chip grid inductance contributes to the noise significantly or not. This is given by the expression:

$$\Delta V_L = L di(t)/dt \tag{1}$$

2.2   IR Noise

The IR noise, $\Delta V_R = i(t) \cdot R$, is due to the on-chip supply grid resistance. The top-most metal layers, which are generally thicker than the lower layers, are mostly allocated to power supply routing. Thus, the resistance of these layers dominates the on-chip grid resistance [7]. The usage of multiple layers for the on-chip grid and multiple parallel supply lines also provides to shrink this resistance to some degree [7]. It is extremely difficult to simulate the entire power supply grid.

3 Budget for Decoupling Capacitance

A common technique to suppress the supply noise is allocation of decoupling capacitors (decap) at appropriate locations. Further, we investigate the voltage drop module–wise, and based on the voltage drop, decap is placed.

The dynamic power consumption can be estimated by the following expression:

$$P = C_T V_{DD}^2 f\gamma. \tag{2}$$

where P is the power consumption, C_T is the chip capacitance, V_{DD} is the supply voltage, f is the clock frequency and γ is the probability of 0–1 transition. In [8], the authors have shown that the value of decap needed to limit the supply noise to less than 10%:

$$C_{decap} = 9P/f V_{DD}^2. \tag{3}$$

From Eq. (3), the total decoupling capacitance can be estimated. This expression has been used at the pre-layout level for noise suppression and analysis [9–11].

In this work, firstly we check the availability of the WS. Secondly, we estimate the supply noise of each module. If the module is placed much nearer to the WS, we place the decap in that particular WS so that there is no or marginal increment in wiring length and area of the IC. If satisfactory reduction of supply noise is not obtained, then value of the decap then can be marginally increased depending on the WS availability. So the supply noise of that module will get reduced without having much effect on the design parameters of the chip. It also increases the possibility of suppressing the supply noise of the nearby modules.

Algorithm: Decap Budget

1. Input: No. of modules, power consumption, operating frequency, supply voltage without decap.
2. Calculate for each module decap budget using Eqn. 3.
3. Check whether satisfactory results are obtained
4. If satisfactory results are obtained
5. Stop
6. Else
7. Change the value of decap marginally
8. Continue step 3 to step 5 unless satisfactory results are obtained.
9. Stop
10. Output: Decap budget
11. End

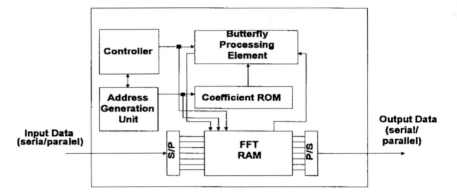

Fig. 1 Architecture of FFT processor [3]

4 CAD Methodology

This paper addresses a CAD approach for decoupling capacitance allocation and hence to suppress noise in a multi-core SoC. It is implemented in TSMC 90 nm technology node. The suppression of supply noise with a minimum increment of delay and power consumption with decap is the main objective of this work. This is completely a new approach in the domain of PDN. The proposed CAD methodology is applicable for any SoC architecture. For analysis, we have considered 512-point FFT architecture as application circuit.

The architecture of FFT processor is shown in Fig. 1 [3]. FFT processors are used in OFDM techniques and modern digital communication systems. In this work, 512-point FFT processor is chosen as it is quite complex and has been used as a test bench circuit.

FFT processor consists of memory, a butterfly unit, controller, and address generator. The butterfly unit is the processing unit which computes the FFT output. ROM is used to store twiddle factors. Input data, intermediate results, and output data are stored in RAM. The control unit is to generate the control signals.

The CAD tools used in this work are as follows:

1. SYNOPSYS (VCS): Verilog code (Functional verification).
2. SYNOPSYS (Design Compiler): (Logic Synthesis).
3. CADENCE (Encounter): (Physical Design).

The Verilog code and functional verification for the FFT processor are done using SYNOPSYS VCS simulator. This design is imported in SYNOPSYS Design Compiler and gate-level netlist is generated. This design is then imported in CADENCE Encounter for physical design, and the design parameters are extracted.

Table 1 Analysis of power with allocation of decap in 90 nm technology node for supply voltage 1 V and operating frequency 1 GHz

FFT size	Power without decap (mw)	Power with decap (mw)	% increment in power
512-point	118.2	121.5	2.71
512-point [11]	159.2	162.5	2.03

Table 2 Noise suppression of FFT cores with allocation of decap in 90 nm technology node for supply voltage 1 V and operating frequency 1 GHz

FFT size	Peak noise without decap (mv)	Peak noise with decap (mv)	% of noise suppression
512-point	0.32	0.2	37.5
512-point [11]	0.62	0.386	37.7

Table 3 Delay analysis of FFT cores with allocation of decap in 90 nm technology node for supply voltage 1 V and operating frequency 1 GHz

FFT size	Without decap delay (μs)	With decap delay (μs)	% increment in delay
512-point	162.4	164.8	1.45
512-point [11]	211.4	213.8	1.12

5 Simulation Results and Analysis

Verilog code for 512-point FFT architecture is written in SYNOPSYS VCS. The design is synthesized in DESIGN COMPILER of SYNOPSYS. The physical design is done in CADENCE ENCOUNTER. Then the supply noise, power consumption, and delay are estimated. Decoupling capacitance is computed using Eq. (2) and is placed near each noisy module. Next, the supply noise, power consumption, and delay analysis have been made with decap. The results are generated using TSMC 90 nm process technology. The results are compared with [11] where the results are generated at the pre-layout stage using 180 nm process technology.

Table 1 shows that the power consumption is increased by 2.71% for 512-point FFT core after allocation of decap at the post-layout stage and is comparable to that of pre-layout stage [11].

In Table 2, it is observed that power supply noise has been suppressed by 37.5% for 512-point FFT core after allocation of decap and is identical to that of pre-layout stage [11].

Table 3 shows that the delay is increased by 1.45% for 512-point FFT core after the proper allocation of decap. The results of pre-layout stage are also shown [11].

From simulation results, it is observed that appropriate decap budget and proper decap placement can reduce the supply noise to a satisfactory level and this can be very helpful for the designers. From simulation results, it can be seen that increment in power consumption, noise suppression, and propagation delay is very much identical to that of pre-layout stage [11].

6 Conclusion and Future Scope

In present VLSI design domain, there are CAD tools that are used for decap budget and allocation. In this work, module-wise approach for proper decap budget and placement to keep the peak noise, delay and power parameters within the satisfactory limits has been proposed. Simulation results show that the decap budget is appropriate and noise is suppressed considerably. Our approach enables a reduction in power supply noise by a significant amount. Also, delay and power increment is within satisfactory limits after placement of decap.

Acknowledgements This work is supported by Advanced VLSI Design Laboratory, IIT Kharagpur.

References

1. S. Zhao, K. Roy, C.K. Koh, Estimation of inductive and resistive switching noise on power supply network in deep sub-micron CMOS circuits, in *Proceedings International Conference on Computer Design*, pp. 65–72 (2000)
2. H. Su, K. Gala, S. Sapatnekar, Fast analysis and optimization of power/ground networks, in *Proceedings of International Conference Computer-Aided Design*, pp. 477–480 (2000)
3. E. Cetin, R.C.S. Morling, I. Kale, An extensible complex fast fourier transform processor chip for real-time spectrum analysis and measurement. IEEE Trans. Instrum. Measure. **47**(2), 95–99 (1998)
4. M. Zhao, R. Panda, S. Sapatnekar, T. Edwards, R. Chaudhry, D. Blaauw, Hierarchical analysis of power distribution networks, in *Proceedings of Design Automation Conference*, pp. 150–155 (2000)
5. J.M. Wang, T. Nguyen, Extended Krylov subspace method for reduced order analysis of linear circuits with multiple sources, in *Proceedings Design Automation Conference*, pp. 247–252 (2000)
6. M. Ang, R. Salem, A. Taylor, An on-chip voltage regulator using switched decoupling capacitors, in *Proceedings International Solid-State Circuits Conference Digital Technical Papers*, pp. 438–439 (2000)
7. A.H. Ajami, K. Banerjee, M. Pedram, Scaling analysis of on-chip power grid voltage variations in nanometer scale ULSI, in *Analog Integrated Circuits and Signal Processing*, vol. 42 (Springer Science Business Media, 2005), pp. 277–290
8. C.S. Chang, A. Oscilowski, R.C. Bracken, Future challenges in electronics packaging. IEEE Trans. Circuits Dev. **14**(2), 45–54 (1998)
9. M. Chakraborty, K. Guha, D. Saha, P. Mitra, A. Chakrabarti, Pre-layout decoupling capacitance estimation and allocation for noise-aware crypto-SoC applications. J. Low Pow. Electron. **11**(3), 333–339 (2015)
10. P. Mitra, C. Roy Chowdhury, Pre-layout noise suppression and delay estimation with decap allocation for 1-k point FFT core, in *Proceedings of 2016 Second International Conference on Research in Computational Intelligence and Communication Network (ICRCICN 2016)*, pp. 247–251 (2016)

11. P. Mitra, J. Bhaumik, Pre-layout decap allocation for noise suppression and performance analysis for 512-point FFT core, in *Proceedings of 2017 Devices for Integrated Circuits (DevIC2017)* (2017)
12. T. Karim, On-chip power supply noise: scaling, suppression and detection, in *Dissertation in Doctor of Philosophy (Electrical and Computer Engineering)* (The University of Waterloo, 2012)
13. S. Pant, Design and analysis of power distribution networks in VLSI circuits, in *Dissertation in Doctor of Philosophy* (The University of Michigan, Electrical Engineering, 2008)
14. K. Shah, *Power Grid Analysis in VLSI Designs: Dissertation in Master of Science (Engineering)* (Super Computer Education and Research Centre, Indian Institute of Science Bangalore, 2007)
15. K. Shimazaki, T. Okumura, A minimum decap allocation technique based on simultaneous switching for nanoscale SoC, in *Proceedings IEEE Custom Integrated Circuits Conference* (2009)

Enhanced Performance of GaN/InGaN Multiple Quantum Well LEDs by Shallow First Well and Stepped Electron-Blocking Layer

Mainak Saha and Abhijit Biswas

Abstract This work analyzes the effect of incorporating a shallow first well and a two-step electron-blocking layer (EBL) on the performance of an InGaN/GaN rectangular multiple quantum well LED using well-calibrated numerical simulator SILVACOATLAS. The results show 78% improvement in output power of the proposed LED structure in comparison with the conventional structure at 150 mA injection current. In addition, the droop in internal quantum efficiency (IQE) for the proposed LED structure reduces to 29% from 54% which is observed in conventional structure. The simulation study reveals that the shallow first well facilitates to reduce the effective blocking provided by the preceding GaN layer to the injected electrons, thereby increasing the electron injection. In addition, the stepped electron-blocking layer raises the energy barrier for electrons at the EBL causing better confinement of electrons compared to conventional structure. The stepped EBL also decreases the effective energy barrier for holes significantly, thereby aiding better injection of holes into the multiple quantum well light-emitting region. Consequently, carrier concentrations in the active region of the proposed structure increase, thereby enhancing its output power at high input current.

Keywords AlGaN electron-blocking layer · Efficiency droop · InGaN/GaN MQWs · Light-emitting diodes · SILVACOATLAS

M. Saha (✉)
Institute of Engineering & Management, Maulana Abul Kalam Azad University of Technology, Salt Lake, Kolkata 700091, India
e-mail: mainaks25@gmail.com

A. Biswas
Institute of Radio Physics and Electronics, University of Calcutta, 92, Acharya Prafulla Chandra Road, Kolkata 700009, India

© Springer Nature Singapore Pte Ltd. 2017
J. Bhaumik et al. (eds.), *Communication, Devices, and Computing*, Lecture Notes in Electrical Engineering 470, https://doi.org/10.1007/978-981-10-8585-7_20

1   Introduction

Nitride-based LEDs have received significant research attention because of their widespread applications in the field of backlighting, liquid crystal display (LCD), mobile platforms, and white solid-state lamps [1–4] since its first demonstration in early 1990s. Despite extensive research efforts toward technological improvement of nitride LEDs and their successful commercialization, the performance of these LEDs is found to be greatly degraded by the drastic drop of internal quantum efficiency when the injection current is increased, termed as efficiency droop [5, 6]. Although the actual reason of the efficiency droop is yet to be figured out, Auger recombination [7], carrier leakage from active region [8], limited hole injection in the device [9] have been proposed as the possible mechanisms responsible for the efficiency droop in InGaN LEDs. Besides, the internal electric field arising from polarization charges in nitride materials is considered to play an important role on efficiency droop [10, 11]. However, the real reason for efficiency droop is still debatable. Several innovative methods providing good reduction in efficiency droop have been proposed, such as AlGaN EBL grown in active region [12], double electron-blocking layer [13], superlattice layer of AlGaN/GaN between n-GaN and first well of active region [14], decreasing barrier width toward anode end [15] step-stage multiple quantum wells and hole-blocking layer [16], and graded quantum wells [17]. From all those reported methods, polarization in the electron-blocking layer can be identified as one major factor causing degradation of the efficiency at high current. The polarization field present in the EBL decreases the effective barrier to electrons while increases the effective barrier to holes. This results in poor electron confinement in the quantum wells and inadequate hole injection in the LED structure [18]. Many techniques of reducing the polarization field in the EBL and thereby enhancing its effectiveness were reported, including graded EBL [11, 18, 19], AlInN EBL for lattice matching [20], AlInGaN EBL for polarization matching [21, 22], inverting EBL polarization by wafer or chip-bonding process [23]. All these methods are either very complex to realize in practice or the results obtained are not satisfactory.

   To mitigate the aforesaid problem, we propose an LED structure endowed with a shallow first quantum well and a two-step electron-blocking layer. The shallow first well helps to lower the polarization field at the boundary of the well with its preceding GaN layer and enhance injection of electrons in the active region. Additionally, the stepped EBL reduces the polarization induced field in the EBL region. This aids to achieve stronger electron blocking and better flow of holes into the active region compared to the conventional LED structure.

2   Device Structure

The conventional LED structure used for the purpose of comparison is an experimentally fabricated device by Kuo et al. [24]. The structure reported in Ref. [24]

Fig. 1 Schematic cross-sectional view of the light-emitting diode structure

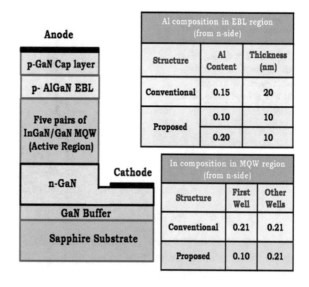

Al composition in EBL region (from n-side)		
Structure	Al Content	Thickness (nm)
Conventional	0.15	20
Proposed	0.10	10
	0.20	10

In composition in MQW region (from n-side)		
Structure	First Well	Other Wells
Conventional	0.21	0.21
Proposed	0.10	0.21

was grown on a substrate of c-plane sapphire, over which a 50 nm-wide undoped GaN buffer layer and a n-GaN layer with a thickness of 4.5 μm and n-doping of 5×10^{18} cm^{-3} was placed. Five In$_{0.21}$Ga$_{0.79}$N (2 nm) quantum wells with six GaN barriers (15 nm) formed the active region. The active region was followed by a p-type $(1.2 \times 10^{18}$ cm$^{-3})$ Al$_{0.15}$Ga$_{0.85}$N electron-blocking layer (EBL) having thickness of 20 nm and a p-type $(1.2 \times 10^{18}$ cm$^{-3})$ GaN cap layer having thickness of 0.5 μm. The proposed LED structure features a shallow first well (counted from n-GaN side) with 10% In content and a stepped electron-blocking layer as illustrated in Fig. 1. The stepped AlGaN EBL is formed by breaking the 20 nm AlGaN EBL into two segments having 10 nm thickness each. The Al composition in the segment next to the last barrier is kept at 10%, and then, it increases to 20% in the following segment. The total Al content in the EBL is kept the same for both the structures. The dimension of the device is considered to be 300×300 μm^2.

3 Simulation Framework

The LED structure is numerically investigated using SILVACOATLAS [25] to obtain the output optical power and internal quantum efficiency variation against input current, the energy band diagram, and also the concentration distributions of electrons and holes. Our simulation program includes coupled Schrodinger and Poisson equations, current continuity equation for both holes and electrons, drift-diffusion equations for transport, and also the photon rate equation. In addition, it incorporates energy band calculation by 6×6 **k.p** model for with carrier escape and flying over phenomena. To model the recombination events along with the radiative process,

Fig. 2 Comparison of our
simulation results with
reported experimental data
[24] for the variation of
output power and voltage
across LED with input
current for the conventional
LED

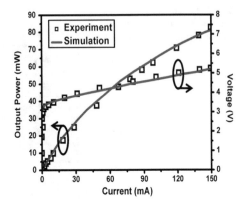

we incorporate non-radiative processes like Shockley-Read-Hall (SRH) and Auger recombination in our simulation following Ref. [26]. For the simulation, the band gap energies of InGaN and AlGaN ternary alloys are calculated as

$$Eg\left(In_yGa_{1-y}N\right) = y \cdot E_g\left(InN\right) + (1 - y) \cdot E_g\left(GaN\right) - 1.43 \cdot y \cdot (1 - y) \quad (1)$$

$$Eg\left(Al_yGa_{1-y}N\right) = y \cdot E_g\left(AlN\right) + (1 - y) \cdot E_g\left(GaN\right) - 0.7 \cdot y \cdot (1 - y) \quad (2)$$

$E_g(GaN)$, $E_g(InN)$, and $E_g(AlN)$ are the band gap energies of GaN, InN, and AlN (at temperature of 300 K) with values 3.435, 0.711, and 6.138 eV, respectively [27]. The offset ratio for energy band is selected to be $\Delta E_C/\Delta E_V = 0.66/0.34$ for both InGaN and AlGaN. The polarization field plays an important role in Group III-nitrides due to large mismatch of electronegativity and lattice constant. The polarization charge densities arising from spontaneous and piezoelectric polarization are taken care of as proposed by Fiorentini et al. [28]. The strain developed in the AlGaN or InGaN alloy grown on GaN base layer is calculated as $\varepsilon = \frac{a_{sub} - a}{a}$, where a_{sub} and a are the lattice constants of the GaN layer and the alloy, respectively. Moreover, we use 40% of polarization charge screening [27] in our numerical analysis.

Figure 2 compares simulated curves of voltage versus injection current (V-I) and optical output versus injection current (L-I) for a conventional LED with their corresponding experimental curves reported in Ref. [24]. A very good consistency is observed between our simulated curves and characteristics reported in Ref. [24] as may be observed from Fig. 2. This validates simulation deck.

4 Results and Discussion

Having calibrated our simulation setup with experimental data [24], we obtain the output optical power and internal quantum efficiency variation against input current for our proposed LED structure as demonstrated in Fig. 3 and Fig. 4, respectively.

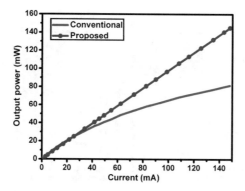

Fig. 3 Output power variation with input current for conventional and proposed LED

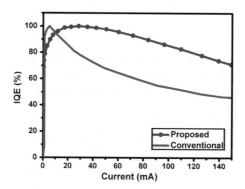

Fig. 4 Internal quantum efficiency variation with input current for conventional and proposed LED

Clearly, Fig. 3 shows that the proposed structure exhibits improved output power compared to conventional structure at same injection current. For example, the proposed structure provides an output power of 145 mW at an injection current of 150 mA, which is 78% more compared to 81.5 mW for the conventional structure. From Fig. 4, the percentages of efficiency droop defined as $\frac{IQE_{peak} - IQE_{150mA}}{IQE_{peak}} \times 100\%$ for the conventional and proposed structures are found to be 54% and 29%, respectively. The results show notable improvement in both output power and IQE for the proposed LED structure compared to the conventional LED structure. To investigate the underlying physical reasons behind this remarkable improvement, the energy band diagram and both electron and hole concentrations for the conventional as well as proposed LED structures are analyzed in detail.

Figure 5 shows the energy band diagram for both conventional and proposed LED structures along with position of quasi-Fermi levels of electrons and holes at an injection current of 120 mA. One may clearly observe that the first well (from n-side) of the proposed LED structure is a shallow well as it has a smaller In composition of 10% compared to other wells with In content of 21%. This leads to less lattice

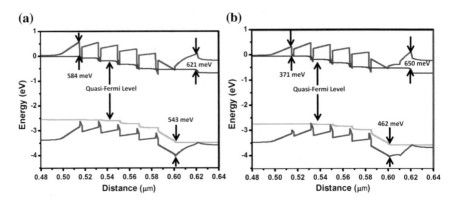

Fig. 5 Energy band diagrams for **a** conventional and **b** proposed LED structures at injection current of 120 mA

mismatch between the first well and the preceding GaN layer and hence creates less amount of electric field resulting in less upward alignment in the conduction band of GaN layer. This feature results in lowering the energy barrier for electrons entering from the n-GaN. The proposed structure offers an effective energy barrier of 371 meV which is much smaller than the effective energy barrier provided by the conventional structure (584 meV) to the electrons entering the active region. So, electrons can easily enter the active region in the proposed structure due to the presence of a shallow first well. The last GaN barrier of the proposed structure is followed by a 10 nm thick $Al_{0.10}Ga_{0.90}N$ instead of the higher Al composition $Al_{0.15}Ga_{0.85}N$ layer of conventional structure. Such a structure also reduces the polarization field present in the EBL of proposed structure. Consequently, the effective energy barrier for electrons in the EBL of proposed LED (650 meV) is found to be more compared to the effective barrier for electrons in the EBL of conventional structure (621 meV). Hence, the proposed LED design has a better electron-blocking capability compared to the conventional structure. The effective barrier height for holes is found to be lower in the proposed design (462 meV) compared to that in the conventional structure (543 meV) due to the less polarization field developed in the EBL region. This helps to achieve a better hole injection into the active region for the proposed LED design.

Figure 6a compares the electron concentration in the active region of the conventional and proposed structure at an input current of 120 mA. It can be seen from the figure that though the electron concentration in the first well of the proposed structure is reduced significantly compared to that of the conventional structure due to the shallow quantum well, the electron concentration increases considerably in all other wells of the proposed structure due to the better hole injection due to the use of shallow first well and better electron confinement provided by the stepped EBL layer. For example, the electron concentration in the last quantum well of proposed structure is almost 20% higher than the electron concentration in the last well of the conventional structure. Figure 6b compares the electron concentration in the EBL region of the conventional and proposed structures in a logarithmic scale. Clearly, it

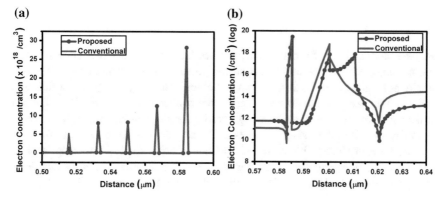

Fig. 6 Distribution of electron concentration in the **a** active region (linear scale) and **b** EBL region (log scale) for the conventional and proposed structure at injection current of 120 mA

Fig. 7 Distribution of hole concentration in the active region (linear scale) for the conventional and proposed structures at injection current of 120 mA

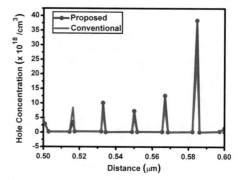

is evident that the electron concentration in the p-GaN region of the proposed LED decreases almost by an order compared to that of the conventional structure. This also suggests that the use of the stepped EBL in the proposed LED blocks the electrons and helps to confine them inside the active region.

Figure 7 compares the distribution of hole concentration in the active region at injection current of 120 mA for both the conventional and proposed structures. The figure suggests that the proposed structure increases the hole concentration in all the quantum wells except the first well due to its lower depth. For instance, the hole concentration in the last quantum well of proposed structure is found to be 50% higher than the hole concentration in the last well of conventional structure. This increase in the hole concentration is attributed to the better hole injection facilitated by the stepped EBL of the proposed structure. The enhanced carrier concentrations in the wells of the proposed structure increase the radiative recombination significantly which results in the improvement of optical performance of the proposed structure.

5 Conclusion

We have studied the performance of a proposed LED structure with shallow first well and stepped electron-blocking layer and compared its optical performance with conventional LED—on the basis of output power and efficiency droop and evaluated in terms of their energy band diagrams and carrier concentration distributions. The proposed LED structure reduces the energy barrier for electron injection into the active region by use of the shallow quantum well. On the other hand, the stepped EBL helps to achieve better electron blocking and injection of holes into the active region. Consequently, the proposed structure offers considerable reduction in efficiency droop that makes it a suitable for implementation of highly luminescent LEDs.

References

1. H.Y. Ryu, K.S. Jeon, M.G. Kang, H.K. Yuh, Y.H. Choi, J.S. Lee, A comparative study of efficiency droop and internal electric field for InGaN blue lighting-emitting diodes on silicon and sapphire substrates. Sci. Rep. **7**, 44814 (2017)
2. T. Jeong, H.-J. Park, J.-W. Ju, H.S. Oh, J.-H. Baek, J.-S. Ha, G.-H. Ryu, H.-Y. Ryu, High efficiency InGaN blue light-emitting diode with >4 W output power at 3 A. IEEE Photon. Technol. Lett. **26**(7), 649–652 (2014)
3. Y.-S. Yoo, J.-H. Na, S.J. Son, Y.-H. Cho, Effective suppression of efficiency droop in GaN-based light-emitting diodes. Role of significant reduction of carrier density and built-in field. Sci. Rep. **6**, 34586 (2016)
4. N. Tansu, H. Zhao, G. Liu, X.-H. Li, J. Zhang, H. Tong, Y.-K. Ee, III-Nitride photonics. IEEE Photon. J. **2**(2), 241–248 (2010)
5. G. Verzellesi, D. Saguatti, M. Meneghini, F. Bertazzi, M. Goano, G. Meneghesso, E. Zanoni, Efficiency droop in InGaN/GaN blue light-emitting diodes: physical mechanisms and remedies. J. Appl. Phys. **114**(7), 71101 (2013)
6. J. Piprek, Efficiency droop in nitride-based light-emitting diodes. Phys. Stat. Sol. (a) **207**(10), 2217–2225 (2010)
7. J. Iveland, L. Martinelli, J. Peretti, J.S. Speck, C. Weisbuch, Direct measurement of Auger electrons emitted from a semiconductor light-emitting diode under electrical injection. Identification of the dominant mechanism for efficiency droop. Phys. Rev. Lett. **110**(17), 177406 (2013)
8. T. Lu, S. Li, C. Liu, K. Zhang, Y. Xu, J. Tong, L. Wu, H. Wang, X. Yang, Y. Yin, G. Xiao, Y. Zhou, Advantages of GaN based light-emitting diodes with a p-InGaN hole reservoir layer. Appl. Phys. Lett. **100**(14), 141106 (2012)
9. Z.G. Ju, W. Liu, Z.-H. Zhang, S.T. Tan, Y. Ji, Z.B. Kyaw, X.L. Zhang, S.P. Lu, Y.P. Zhang, B.B. Zhu, N. Hasanov, X.W. Sun, H.V. Demir, Improved hole distribution in InGaN/GaN light-emitting diodes with graded thickness quantum barriers. Appl. Phys. Lett. **102**(24), 243504 (2013)
10. H. Zheng, H. Sun, M. Yang, J. Cai, X. Li, H. Sun, C. Zhang, X. Fan, Z. Zhang, Z. Guo, Effect of polarization field and nonradiative recombination lifetime on the performance improvement of step stage InGaN/GaN multiple quantum well LEDs. J. Disp. Technol. **11**(9), 776–782 (2015)
11. Y.-K. Kuo, J.-Y. Chang, M.-C. Tsai, Enhancement in hole-injection efficiency of blue InGaN light-emitting diodes from reduced polarization by some specific designs for the electron blocking layer. Opt. Lett. **35**(19), 3285–3287 (2010)

12. Z. Zhang, H. Sun, X. Li, H. Sun, C. Zhang, X. Fan, Z. Guo, Performance enhancement of blue light-emitting diodes with an undoped AlGaN electron-blocking layer in the active region. J. Disp. Technol. **12**(6), 573–576 (2016)

13. Y. Guo, M. Liang, J. Fu, Z. Liu, X. Yi, J. Wang, G. Wang, J. Li, Enhancing the performance of blue GaN-based light emitting diodes with double electron blocking layers. AIP Adv. **5**(3), 37131 (2015)

14. M. Saha, A. Biswas, Studies on reduction of efficiency droop in InGaN/GaN multiple quantum well LEDs. INROADS (Special Issue) **3**(1), 225–229 (2014)

15. H. Karan, A. Biswas, M. Saha, Improved performance of InGaN/GaN MQW LEDs with trapezoidal wells and gradually thinned barrier layers towards anode. Opt. Commun. **400**, 89–95 (2017)

16. Z. Zheng, Z. Chen, Y. Chen, H. Wu, S. Huang, B. Fan, Z. Wu, G. Wang, H. Jiang, Improved carrier injection and efficiency droop in InGaN/GaN light-emitting diodes with step-stage multiple-quantum-well structure and hole-blocking barriers. Appl. Phys. Lett. **102**(24), 241108 (2013)

17. A. Salhi, M. Alanzi, B. Alonazi, Effect of the quantum-well shape on the performance of InGaN-based light-emitting diodes emitting in the 400–500 nm range. J. Disp. Technol. **11**(3), 217–222 (2015)

18. C.H. Wang, C.C. Ke, C.Y. Lee, S.P. Chang, W.T. Chang, J.C. Li, Z.Y. Li, H.C. Yang, H.C. Kuo, T.C. Lu, S.C. Wang, Hole injection and efficiency droop improvement in InGaN/GaN light-emitting diodes by band-engineered electron blocking layer. Appl. Phys. Lett. **97**(26), 261103 (2010)

19. N. Zhang, Z. Liu, T. Wei, L. Zhang, X. Wei, X. Wang, H. Lu, J. Li, J. Wang, Effect of the graded electron blocking layer on the emission properties of GaN-based green light-emitting diodes. Appl. Phys. Lett. **100**(5), 53504 (2012)

20. S. Choi, H.J. Kim, S.-S. Kim, J. Liu, J. Kim, J.-H. Ryou, R.D. Dupuis, A.M. Fischer, F.A. Ponce, Improvement of peak quantum efficiency and efficiency droop in III-nitride visible light-emitting diodes with an InAlN electron-blocking layer. Appl. Phys. Lett. **96**(22), 221105 (2010)

21. M.F. Schubert, J. Xu, J.K. Kim, E.F. Schubert, M.H. Kim, S. Yoon, S.M. Lee, C. Sone, T. Sakong, Y. Park, Polarization-matched GaInN/AlGaInN multi-quantum-well light-emitting diodes with reduced efficiency droop. Appl. Phys. Lett. **93**(4), 41102 (2008)

22. A.J. Ghazai, S.M. Thahab, H. Abu Hassan, Z. Hassan, Quaternary ultraviolet AlInGaN MQW laser diode performance using quaternary AlInGaN electron blocking layer. Opt. Express **19**(10), 9245–9254 (2011)

23. D.S. Meyaard, G.-B. Lin, M. Ma, J. Cho, E. Fred Schubert, S.-H. Han, M.-H. Kim, H. Shim, Y. Sun Kim, GaInN light-emitting diodes using separate epitaxial growth for the p-type region to attain polarization-inverted electron-blocking layer, reduced electron leakage, and improved hole injection. Appl. Phys. Lett. **103**(20), 201112 (2013)

24. Y.-K. Kuo, J.-Y. Chang, M.-C. Tsai, S.-H. Yen, Advantages of blue InGaN multiple-quantum well light-emitting diodes with InGaN barriers. Appl. Phys. Lett. **95**(1), 11116 (2009)

25. ATLAS User's Manual, software version 5.18.3.R, Silvaco International, Santa Clara, CA, (2012)

26. Y.-K. Kuo, S.-H. Horng, S.-H. Yen, M.-C. Tsai, M.-F. Huang, Effect of polarization state on optical properties of blue-violet InGaN light-emitting diodes. Appl. Phys. A **98**(3), 509–515 (2010)

27. Y.-K. Kuo, M.-C. Tsai, S.-H. Yen, T.-C. Hsu, Y.-J. Shen, Effect of P-type last barrier on efficiency droop of blue InGaN light-emitting diodes. IEEE J. Quant. Electron. **46**(8), 1214–1220 (2010)

28. V. Fiorentini, F. Bernardini, O. Ambacher, Evidence for nonlinear macroscopic polarization in III–V nitride alloy heterostructures. Appl. Phys. Lett. **80**(7), 1204–1206 (2002)

A μ-Controller-Based Biomedical Device to Measure EMG Strength from Human Muscle

Arindam Chatterjee

Abstract EMG signal is the heart of different muscular activities. The strength of this signal shows the respective muscle strength and identifies if any type of muscles fatigue or disorder is present or not. This is useful for sports personnel's and especially for hand amputation case having different level of amputation. In this paper, we propose a portable hand carrying device that can indicate the strength as well as display the muscle strength in terms of analog RMS value of the collected electromyography signal on a proposed embedded trainer board in terms of "volt" as well as visualize the strength of the signal using strip of LED bars. An EMG signal processing set up, consists of sensor materials and signal processing circuitry, that are used to collect and process the signal in to its proper shape. The processed signal is fed to a centralised processor/controller for further action.

Keywords EMG electrodes · Preamplifier · Filter · RMS to DC
LCD system · μ-controller · Shift register · LED strips · Muscle strength

1 Introduction

Electromyogram (EMG) signal is generated when muscles contract or relax [1, 2] and processed through biological neurons and muscle fibers to perform different task. It has widespread applications in sports and prosthetic industry. Biologically, the strength of the EMG signals is proportional to the applied muscle force. So the strength estimation is an important factor. For a prosthetic surgeon, it is important to identify the muscle strength of amputee's stump, which plays a significant role to operate the arm with different grip forces [3–6]. So the signal produced must be of sufficient strength to activate the control system, and their levels should be such that even if a child can perform repeated contractions without tiring. The key issues of portable biomedical instrumentation are: low power consumption, long-term sen-

A. Chatterjee (✉)
CSIR-Central Scientific Instruments Organization, Chandigarh, Union Territory, India
e-mail: arindamchatterjee@csio.res.in

© Springer Nature Singapore Pte Ltd. 2017
J. Bhaumik et al. (eds.), *Communication, Devices, and Computing*, Lecture Notes
in Electrical Engineering 470, https://doi.org/10.1007/978-981-10-8585-7_21

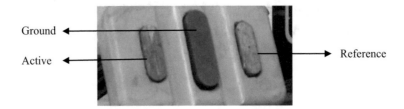

Fig. 1 Titanium-made sensors placed in the skin to receive EMG signal

sor stability, comfortable wearing and wireless connectivity, and low cost. Several low-cost solutions related to EMG processing, signal acquisition, strength estimation using PC-based LAB-View platform [7], and wireless data transfer module like Zigbee, GPRS, Internet [8–11] have been developed so far. In our proposed design, no PC and DAQ board is required for signal acquisition and display, because 2 line LCD is used to display the strength of the RMS value of EMG signal in terms of "volt". The kit can be useful for both the surgeon and amputees to understand the muscle strength in terms of voltage. If the muscle strength is not sufficient, then the provision is there to boost up the signal strength as well.

2 System Description

The overall system can be divided into three following parts.

2.1 EMG Sensors

EMG sensors are made of biocompatible material *titanium* of concentration above 98% with proper shape and size. Three electrodes are used here marked as active, reference, and ground as shown in Fig. 1. These electrodes tap the EMG signal from amputees stump muscle, which conditioned through the signal processing circuitry placed in same casing and just beneath the electrode sets.

2.2 Signal Processing Circuitry

The raw EMG contains noise and different motion artifacts in it. So for removing noise and artifacts, several methodologies have been adopted and their comparison is presented to identify the correct techniques of analysis [12, 13]. The processing circuit is divided into several building blocks as shown in Fig. 2 which are as follows

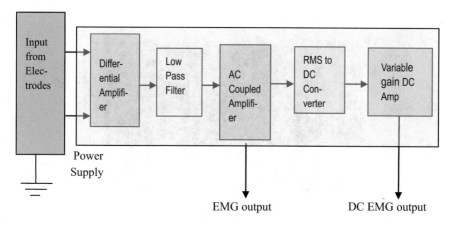

Fig. 2 Block diagram for signal acquisition

Differential Amplifier: At the very first stage, the raw signal which is tapped and fed directly to the preamplifier operates using differential mode having a very high contamination of noise which is mixed up badly with original signal and the signals amplitude is very low, some tenth of microvolt. In this situation, an amplifier has to be designed that have a very high input impedance, high gain, low input noise, high CMRR, and adequate bandwidth. In our system we use AD620 as pre-amplifier IC, whose input impedance is 10GΩ The IC has a special gain resistance, which is variable and should be set as per the requirement. We set the gain as 1000 as per our requirement.

Low-Pass Filter: After the amplification, it has been observed that the frequency components of SEMG signal vary due to contraction and relaxation of arm muscle and most of the signal energy concentrated in the low frequency ranges within 100 Hz. The signals are, therefore, low-pass filtered in this range only, thus rejecting the irrelevant frequencies. The value of "R" and "C" for the "active low-pass filter circuit" has been chosen, so that we can get the value as below

$$f_c = 1/2\pi RC = 1/2\pi * \left(56 \times 10^3\right) * \left(0.1 \times 10^{-6}\right) = 28.42\,\text{Hz}$$

AC-Coupled Amplifier: After passing the signal through filter circuit, noise reduces but due to attenuation caused by the attenuation band of the frequency response of the filter, signal strength decreases. AC-coupled high gain amplifier not only boosts up the signal up to a certain scale, but also due to its capacitive coupling, DC components are blocked if present. AC-coupled amplifier amplifies the signal by a factor of 10,000.

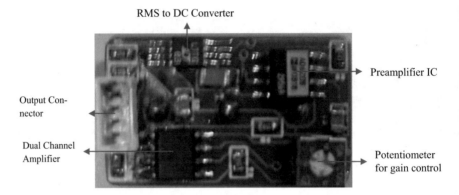

Fig. 3 Total layout of analog signal processing circuitry

RMS-to-DC Converter: This signal is averaged by an RMS-to-DC converter, because a random signal cannot be analyzed properly. The output level of the converter is low in relaxed state, and it increases while force is applied. Thus, it has been seen that the DC level at the output of RMS-to-DC converter varies from 0.8 V to about 4.6 V maximum.

Variable Gain DC amplifier: As the strength of the EMG signal varies from person to person, a variable gain DC amplifier is necessary after RMS to DC where a potentiometer is used to control the overall gain of the circuit. To bring uniformity in the level of DC signal output for different persons, a variable gain DC amplifier is provided, where a potentiometer is used to control the overall gain of the circuit.

Figure 3 shows the image of the designed circuitry composed of SMD IC's. One dual channel amplifier IC is used to fulfill the requirement of AC-coupled amplifier as well as variable gain DC amplifier. Output connector shown is a four-pin connector, one is the power supply input from battery source to IC's, one is actual EMG output after AC-coupled amplifier, one is the DC EMG output from DC variable gain amplifier, and last one is used for overall ground of the board.

2.3 Embedded Circuit Board

Figure 4 shows the block diagram of the trainer board consists of a low-power CMOS 8-bit AVR μ-controller, based on the RISC architecture, which executes powerful instructions in a single clock cycle, 1 MIPS per MHz allowing the system designed to optimize power consumption versus processing speed [14].

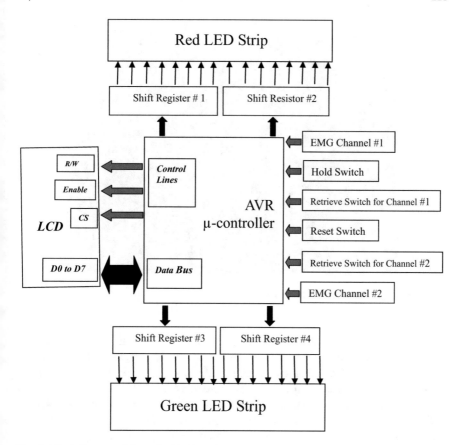

Fig. 4 Block diagram of the embedded board

Four serial in parallel out (SIPO)-type shift registers are interfaced with μ-controller, which are connected with a strip of 32 LEDs. It receives the clock signal from μ-controller unit and generates parallel outputs that glow the LEDs sequentially according to the strength of the EMG signal. Two different strips of LEDs are taken to show the strength of the signal acquired from two different channels. Apart from that another two-line dot matrix, LCD is interfaced with μ-controller that can display up to one 8-character line or two 8-character lines. It displays the text related to the strength of EMG signal in the form of DC voltages.

The low power supply (2.7–5.5 V) of the LCD is suitable for any portable battery-driven product requiring low power dissipation. The backlight provisions are also present (anode and cathode).

Fig. 5 Photograph of embedded board

The display has three main control pins (R/W, CS, and Enable). These three pins are connected directly with μ-controller as instruction pin, and an 8-bit data bus D0–D7 is interfaced as a bidirectional data bus. The control pins are used for synchronization with μ-controller unit. Data are transferred bidirectionally by setting and resetting the control pins properly. Three μ-switches named as "Hold," "Reset," and "Retrieve" are also interfaced with μ-controller unit. These are used to hold the values of input signals for proper operation. Figure 5 shows the photograph of embedded trainer kit.

3 System Operation

Electromyogram is the electrical signal generated by muscles during contraction or relaxation. This electrical activity comes from the muscle itself, without using any shock therapy. Two EMG electrodes are used to acquire the electromyogram signal from two antagonistic muscles. *Anconeus, Pronator-teres, Brachioradialis* are identified as antagonistic muscles for below elbow amputation, where as *biceps* and *triceps* are identified as antagonistic muscles for above elbow amputation. After system initialisation "Retrieve" button is pressed to acquire DC-EMG signal from two channels, and wait for sometimes until the signals are stable. Then, the "Hold" button is pressed to acquire those signals. LCD and LED are used to show its strength. Figs. 6 and 7 shows the profile of the EMG signal collected at the time of relaxation and contraction of muscles.

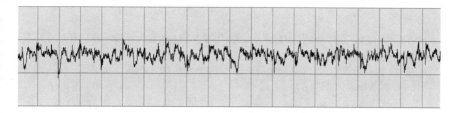

Fig. 6 Relaxation of muscle

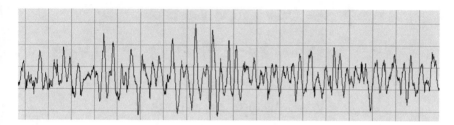

Fig. 7 Contraction of muscle

Fig. 8 Old photographs taken in our laboratory

Figure 8 shows some old photographs of the experiment done in our lab. The electrodes were made of Ag-AgCl materials. A chemical solution "Ten20" was used to place on the surfaces of electrodes before use. This paste creates a very good conductivity among the electrode surface and skin surface. All the boards were hand soldered and the IC packages were DIP. There is no LCD, only LED strips were used to identify muscle strength. The experiment was organised for proof of concept.

4 Calculation

Suppose we have "N" numbers of LEDs. The voltage levels are identified as (V_1, V_2, V_3,...V_N) sequentially.

V_{max} = maximum voltage to glow "N" numbers of LEDs at a time. So we can define the voltages as follow

$$V_1 = \frac{V_{max}}{N}; V_1 => \text{Voltage to glow one LED}$$

$$V_2 = 2\frac{V_{max}}{N}; V_2 => \text{Voltage to glow two LED's}$$

$$V_3 = 3\frac{V_{max}}{N}; V_3 => \text{Voltage to glow three LED's}$$

$$V_{N-2} = \frac{N-2}{N} * V_{max}; V_{N-2} => \text{Voltage to glow } (N-2) \text{ LED's}$$

$$V_{N-1} = \frac{N-1}{N} * V_{max}; V_{N-1} => \text{Voltage to glow } (N-1) \text{ LED's}$$

$$V_N = V_{max}; \quad V_N => \text{Voltage to glow all N LED's.}$$

In our case, we assume $N = 16$ (16 LEDs we have used for our experiment). While collecting the data from amputees stump, we received $V_{max} = 4.6$ V from an amputee's stump.

As the total number of LED's were 16. So we divide the maximum voltage with 16 and get the value to glow one LED, that will be classified as minimum muscle strength. Next by using aforementioned relationship, we will be able to calculate all the strengths to glow all the LED's one by one. The calculated values are as

$$V_1 \sim 0.28\,\text{V}; V_2 \sim 0.56\,\text{V}; V_3 \sim 0.84\,\text{V}; V_4 \sim 1.12\,\text{V}; V_5 \sim 1.40\,\text{V};$$
$$V_6 \sim 1.68\,\text{V}; V_7 \sim 1.96\,\text{V} \ldots V_{16} = 4.6\,\text{V}.$$

5 Flow Chart Diagram

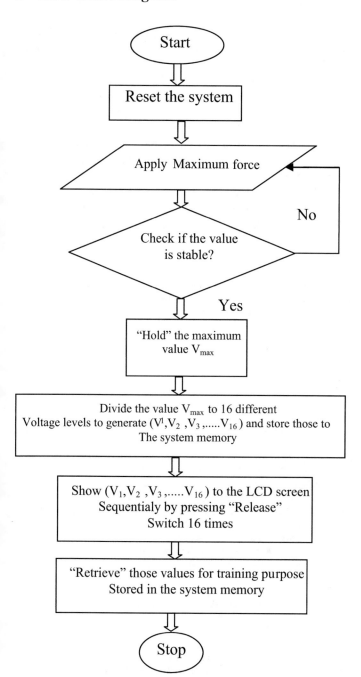

6 Result and Discussion

As stated above, the portable device is useful to show the muscle signal strength from amputees before rehabilitation. The author has done all the experiment himself by putting the sensors in his hand muscle as shown in the photograph and collected the signal to check whether the system is working or not. Later on, the electrode materials were changed into "titanium," which is a biocompatible and more efficient than Ag–AgCl electrode sets. The electrodes used here are totally noninvasive in nature. Signal is collected from skin surface, which never results any gangling or swelling, because the electrodes were made using body compatible materials titanium. It is not implanted through any invasive surgery. As per norm and regulation of medical instrument, noninvasive system does not require any ethical clearance. If it is invasive, then and only then ethical clearance is required. Afterward, it has been experimented with amputee stumps with the consent of orthopedic surgeon at NIOH, Kolkata; AIIMS, Delhi; ALIMCO, Kanpur; before the fitments of myoelectric hand prosthesis. We collected the data from patient with that device, but that cannot be disclosed officially due to government policy.

Acknowledgements I, Arindam Chatterjee, expressed my sincere thanks to Mr. Niranjan V. K. and Mr. Prakhar Aggarwal who helped me to make the overall setup and experiment and to Director CSIR-CSIO for allowing me to submit this article to conference proceedings.

References

1. J.H. Quach, *Surface Electromyography: Use, Design & Technological Overview* (Concordia University, 2007), ID: 4699483
2. C.J. De Luca, *Surface Electromyography: Detection and Recording* (Del-Sys Incorporated, 2002)
3. Z.O. Khokhar, Z.G. Xiao, C. Menon, Khokhar, Surface EMG pattern recognition for real-time control of a wrist exoskeleton. Bio-Med. Eng. On Line **9**, 41 (2010), http://www.biomedical-engineering-online.com/content/9/1/41
4. M. Zecca, S. Micera, M.C. Carrozza, P. Dario, Control of multifunctional prosthetic hands by processing the electro-myo graphic signal. Crit. Rev. Biomed. Eng. **30**(4–6), 459–485 (2002)
5. S. Kumar, A. Chatterjee, A. Kumar, Design of a below elbow myo-electric arm with proportional grip force. J. Sci. Ind. Res. (JSIR) **71**, 262–265 (2012)
6. S.N. Siddek, A.J.H. Mohideen, Mapping of EMG signal to hand grip force at varying wrist angles, in *IEEE EMBS International Conference on Biomedical Engineering and Sciences* (2012), pp. 648–653
7. A. Chatterjee, S. Gupta, S. Kumar, K. Garg, A. Kumar, An innovative device for instant measurement of surface electro-myography for clinical use. Measurement **45**(7), 1893–1901 (2012)
8. K. Rendek, M. Darícek, E. Vavrinsky, M. Donoval, D. Donoval, Biomedical signal amplifier for EMG wireless sensor system, in *IEEE Conference*. https://doi.org/10.1109/asdam.2010.5667015
9. P. Bifulco, M. Cesarelli, A. Fratini, M. Ruffo, G. Pasquariello, G. Gargiulo, A wearable device for recording of biopotentials and body movements, in *IEEE International Workshop on Medical Measurement and Application Proceedings*, pp. 469–472. https://doi.org/10.1109/memea.2011.5966735

10. D.K.N. Silva, R.M.V. Sato, A.L.S. Castro, A portable and low cost solution for EMG using ZigBee, GPRS and Internet to biomedical application, in *19th IMEKO Symposium and 17th IWADC Workshop on Advances in Instrumentation and Sensors Interoperability, 18–19 July 2013, Barcelona, Spain* (2013), pp. 81–86

11. K.E. Pramudita, F. Budi Setiawan, Siswanto, Interface and display of electromyography signal wireless measurement, in *1st International Conference on Information Technology, Computer and Electrical Engineering (ICIT ACEE)* (2014), pp. 58–62

12. C. Liu, X. Wang, Development of the system to detect and process electromyogram signals, in *Proceedings of the IEEE Engineering in Medicine and Biology 27th Annual Conference Shanghai, China, 1–4 September 2005*, pp. 6627–6630. https://doi.org/10.1109/iembs.2005.1616021

13. R.H. Chowdhury, M.B.I. Reaz, M.A.B.M. Ali, A.A.A. Bakar, K. Chellappan, T.G. Chang, Surface electromyography signal processing and classification techniques sensors. 12431–12466 (2013). https://doi.org/10.3390/s130912431

14. Data Sheet of Atmega 32, www.atmel.com/Images/doc2503.pdf-UnitedStates

Design and Implementation of a DCM Flyback Converter with Self-biased and Over-Current Protection Circuit

R. Rashmi and M. D. Uplane

Abstract This paper presents a low-cost design of flyback converter in high-voltage and low-power application. The flyback converter is designed with the E–I Ferrite core MTC transformers with high permeability and high saturation point. The control circuit is designed around the integrated chip of UC3844 fixed frequency current mode controller incorporated with error amplifier, current sensing comparator, and a high totem pole output driver. External biasing is not required for the control circuit as it is directly biased from the high-voltage line and high-frequency transformer. Isolation is provided by transformer and optocoupler for input–output and power control circuit, respectively. Voltage and current are sensed through the resister divider circuit and shunt resistor, respectively. Control characteristics and performance of the flyback converter are simulated using PSIM software, and 20 W flyback converter hardwire circuit is built to validate the analytical results under variable load current and variable source voltage. Performance of the converter with variable input voltage (25–55 V) and variable load (50–200 Ω) is reported.

Keywords Multiple output flyback · PWM · Steady state response · Current control loop

1 Introduction

With advancing technology, efficient regulation and power density are in demand in various applications like battery charge, LED driver, photovoltaic cell high-voltage dc transmission, and many other applications [1, 2]. In all these applications, high current and voltage protection is required [3, 4]. These demands can only be fulfilled with high-frequency converters with the shrink size of transformer. So here ferrite core E and I individual laminations are tightly butted together during the inductors construction to reduce the reluctance of the air gap at the joints producing a highly saturated magnetic flux density for efficient work at high frequency [5]. High power

R. Rashmi (✉) · M. D. Uplane
Department of Instrumentation Science, Savitribai Phule Pune University, Pune, India
e-mail: ruchirashmi@gmail.com

© Springer Nature Singapore Pte Ltd. 2017
J. Bhaumik et al. (eds.), *Communication, Devices, and Computing*, Lecture Notes in Electrical Engineering 470, https://doi.org/10.1007/978-981-10-8585-7_22

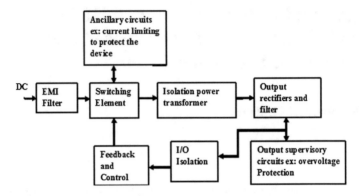

Fig. 1 Block diagram of flyback converter

density and low cost are achieved by the flyback converter as it is based on the primary buck-boost converter topology with input–output isolation and are widely used without output filter inductor and freewheeling diode [6].

The basic block diagram of the designed circuit is shown in Fig. 1. The most important factor in high-frequency converters is safety which is maintained by the isolated feedback and coupled inductor. The power and control circuit isolation is maintained by the optical isolation circuit which provides protection from the potentially lethal voltages and currents as well as maintains the converter accuracy by interrupting the ground loop. It provides better tracking of output voltage with line and load regulation as compared to forward converter. With this magnetic integration, the most cost-effective, simple, and stabilized DC output power supply is designed.

It normally operates in two modes based on the amount of energy transferred from transformer to the load during each cycle. In discontinuous mode or complete energy transfer mode, all the energy of the transformer stored during the switch on period is completely transferred to the output during the switch-off period while in continuous mode or incomplete energy transfer mode, some part of energy of the transformer remains in the transformer even during next switch-on period [7–9]. In this paper, the discontinuous mode of flyback is designed and analyzed as it reduces the core size. Comparison of different configuration of topology with the number of devices and isolation is shown in Table 1. Many standard integrated circuits are incorporated with different types of controllers and MOSFET for offline flyback converter as shown in Table 2.

This paper is organized as follows. In Sect. 2, operation principle of the flyback converter is described. Design of different sections of flyback converter with current feedback loop is mentioned in Sect. 3. The result is discussed in Sect. 4. The conclusion is given in Sect. 5.

Table 1 Comparison of some converters

Topology configuration	No. of devices					Isol.
	S	D	L	C	T	
Buck	1	1	1	2	5	No
Synchronous buck	2	0	1	2	5	No
Boost	1	1	1	2	5	No
Inverting buck-boost	1	1	1	2	5	No
Fly-buck	2	1	1	2	5	Yes
Flyback	1	1	1	2	5	Yes
Two-switch flyback	2	3	1	2	8	Yes

S Switches, D Diodes, L Magnetic components (inductors and coupled inductors), C Capacitor, T Total components, *Isol.* High-frequency isolation (Galvanic isolation)

Table 2 Comparison of power integration ICs

Power intergration IC	Switches	Controllers	Diodes	Light or no load performance	Power range (W)	Power topology
InnoSwitch family	1	2	0	No	0–25	Offline flyback flux link technology with isolation
LinkSwitch family	1	1	0	Yes	0–150	Offline flyback with on/off controller without isolation
TOPSwitch family	1	1	0	Yes	0–200	Multimode control
Hiper family	1	1	1	Yes	75–400	CCM boost PFC, two-switch forward and resonant half bridge (LLC)
TinySwitch family	1	1	0	Yes	25	Offline flyback with on/off controller and without isolation

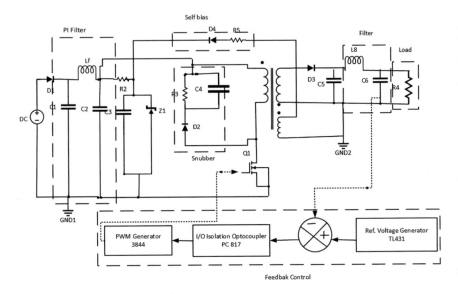

Fig. 2 Configuration of flyback converter with feedback

2 Operation of Flyback Converter

Flyback converter and its feedback control loop are shown in Fig. 2. Here, PI filter is used to minimize the input voltage ripple caused by the discontinuous input current of a step-down regulator. The function of self-bias block is to provide a voltage V_d by the high-frequency transformer after the energization of the transformer higher than the V_c provided through R_2 and Z_1. The snubber circuit RCD reduces the amplitude of the leakage inductance voltage spike during the turn off interval of the switch [10]. The control circuitry comprises of voltage sensor signals, current sensor signals, feedback isolation, and PWM control circuitry.

There are the two operating states of the flyback converter. One is switch-on period (DTs period) or magnetizing period as shown in Fig. 3 when the inductor energy builds up while the capacitor supplies the load current. Second is the switch-off ($(1 - D)$Ts period) period when the inductor energy charges output capacitor and discharges to the load. Figure 4 shows the operative circuit during the $(1 - D)$Ts period.

2.1 Operation During on Period of Switch

During this time, the switch Q_1 is on, the dot ends of all the winding are positive with respect to their non-dot ends. Output rectifier diode D_3 is reverse-biased, and all the output currents are supplied from storage capacitor C_6. The energy builds up

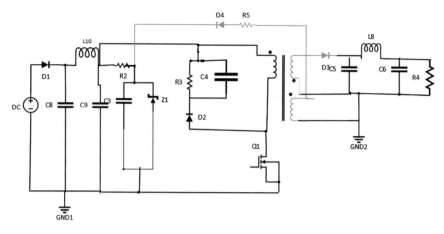

Fig. 3 Operation during the switch-on period

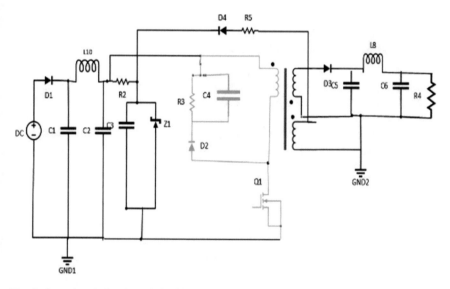

Fig. 4 Operation during the switch-off period

at the primary winding of the transformer as it is directly connected across the DC source input V_{dc} and the current ramps up linearly at the rate of

$$dI/dt = (V_{dc} - 1)/L_p \tag{1}$$

where L_p is the primary inductance. The maximum ramped up primary current is $I_p = V_{dc} \cdot T_{on}/L_p$ at the end of the on time. This current represents a stored energy of

$$E = \frac{Lp \cdot (Ip)^2}{2} \qquad (2)$$

2.2 Operation During off Period of Switch

Figure 4 shows the operative circuit during the $(1 - D)Ts$ period. During this time, the switch Q_1 is off and the diode D_3 is forward-biased. The inductive current forces reversal of polarities on all winding and the non-dot ends become positive with respect to the dot end in primary winding. The primary current transfers to the secondary at an amplitude Is = Ip (Np/Ns). The inductor current ramps down linearly at a rate dIs/dt = Vo/Ls, where Ls is the secondary inductance. The primary winding of the inductor is out of action. The secondary current ramps down to zero before the start of next switch-on time to deliver the stored energy to capacitor and the load.

3 Design of Flyback Converter

Normally flyback converter is used for high voltage and relatively low power in the range of 5–150 W [11]. The flyback was designed with the following specifications as given in Table 3.

The different sections of the converter are integrated to make a complete flyback converter system as shown in Fig. 5. The designed part of the different sections is mentioned below.

Table 3 Design parameters of flyback converter

Design parameters	Values	Specification of transformer winding with number of turns and diameter width		
Vo	21.6 V	Windings	Turns	Diameter (SWG)
Po(max)	20 W	NP_1	30	23
Io(max)	900 mA	NS_1	18	23 * 2
Io(min)	0.1 A	NS_2 (biasing)	10	30
Vdc max	55 V	SWG-standard wire gauge, core geometry-EE core, transformer design-MTC		
Vdc min	25 V			
Switching frequency	40 kHz	Specification of snubber circuit		
Isolated/Non-isolated	Isolated	R	10 kΩ, ½ W	
Topology	Flyback-DCM	C	0.1 μF, 1000 V	

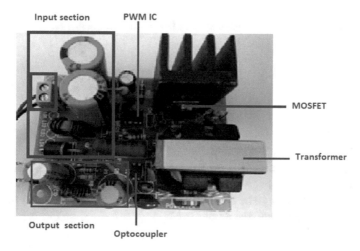

Fig. 5 Hardwired flyback converter

3.1 Input

The input capacitor selected here is based on the RMS current and voltage rating. The selected ceramic capacitor is able to sustain the RMS current without heating.

3.2 Buck Converter

Maximum stress on the MOSFET during off state is 300 V but for the safer side, MOSFET of high-voltage rating is chosen to have less effect of the parasitic ringing. The ratio of the turns of the transformer is calculated as [6].

$$Vomax = Vdcmax + \frac{Np}{Ns}(V0 + 1)$$

$$300 = 90 + \frac{Np}{Ns}(20 + 1)$$

$$\frac{Np}{Ns} = \frac{210}{21} = 10 \tag{3}$$

$$\text{Maximum on time} \qquad Ton = \frac{(V0 + 1)\frac{Np}{Ns}(0.8T)}{(Vdc - 1) + (Vo + 1)\frac{Np}{Ns}} \tag{4}$$

The chosen components based on the above calculation are

MOSFET rating = 8 A, 500 V
Ton = 13.77 μs

Primary inductor calculation is based on the switching frequency and power requirement.

$Lp = 232 \, \mu H$

Peak current $Ip = 2.074$ A

3.3 Snubber Circuit

The designed snubber circuit is an RCD circuit to absorb energy of the leakage inductance. With 40 kHz of switching frequency, the reflected voltage from the primary transformer cannot be clamped straight to the input voltage, so a snubber is used to clamp the energy to the input.

3.4 Feedback Isolation

The optocoupler is used for the signal transmission between circuits of different potentials and impedances. It can give up to voltage isolation of 5000 V. Its response time is 20 μs. For supply voltages ranging from 25 to 55 V, the circuit services a 900 mA load and maintains 0.1% line regulation.

3.5 Control Section

TL 431 shunt regulator compares the feedback signal of the output voltage with the fixed reference voltage, and difference of voltage (error) is sent to the IC UC3844 through optocoupler. When the switch current is lesser than the threshold value switch conducts, and as the switch current reaches the threshold value the conduction stops.

3.6 Output Section

At flyback converter output, only diode and capacitors are required to produce the output as the transformer also acts as an energy-storing inductor. But here, an additional LC filter is used to suppress the high switching spikes.

4   Result and Discussion

The converter is simulated as well as tested on the hardwire. The IC 3844 is control-ling the desired voltage 21.6 with 900 mA load as shown in Fig. 6.

IC 3844 operates in current mode and it senses the switch current through the sense resister, and when the switch current reaches the threshold value established by the shunt regulator by forcing the optocoupler to conduct then the switch conduction stops and again the conduction starts by the 40 kHz oscillator. Fig. 7 shows the

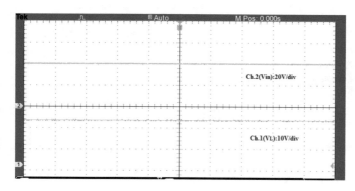

Fig. 6   Regulated voltage

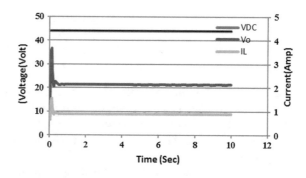

Fig. 7   Input DC voltage, regulated load voltage, and load current response with the time

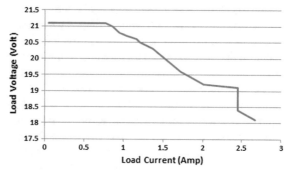

Fig. 8   Variation of load voltage with load current

Fig. 9 Variation of load
voltage with change in input
voltage

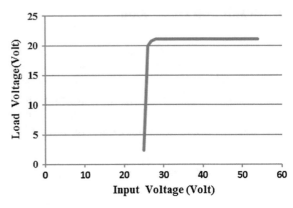

Fig. 10 Variation in load
and input current with
changing load

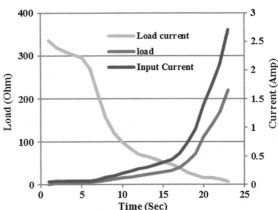

simulation result of the converter where the input DC voltage is fixed at 44 V and
the load DC voltage is 21.6 V and the load current is 0.3 A. At steady state, the
response of the converter is stable with little ripple but the transient response of the
converter shows some peak due to the slow response of the inductive and capacitive
components.

The variation of the load voltage with the maximum load current is shown in Fig. 8.
As the load current increases than 0.9 A, the load voltage becomes unregulated. As
the load current increases beyond designed value, the primary transformer is not able
to transfer the required power to drive the load causes decrease in load voltage to
maintain the load current. The variation of the load voltage with the variation of the
input DC voltage is shown in Fig. 9. At the 25 V DC input, the load voltage is only
2.35 V as the converter is not able to drive the load. For the input voltage of 28 V and
greater, the load voltage is well regulated at set point of 21.6 V. The variation of the
input current load current with the variation of the load and time is shown in Fig. 10.
The load resistance is varied from 7.3 to 335 Ω, input current is varied from 0.042
to 1.655 A, and load current is varied from 0.0612 to 2.71 A. Input current increases
to drive the higher load current with the decrease of resistances.

5 Conclusion

This paper has analyzed and presented a detailed design and a low-power high-voltage flyback converter with feedback isolation. The key components of the design of switch mode power supply like current sensor, voltage sensor, feedback loop with shunt regulator and optocoupler, coupled inductor, controller IC have been described. Low cost and compact size power regulation have been achieved with the integrated current mode PWM controller and minimum number of components.

Acknowledgements This work has been supported by DST-PURSE, Savitribai Phule Pune University. R. Rashmi is thankful to University Grant Commission for SRF.

References

1. J.P. Hong, G.W. Moon, A digitally controlled soft valley change technique for a flyback converter. IEEE Trans. Ind. Electron. **62**(2), 966–971 (2015). https://doi.org/10.1109/TIE.2014.2352600
2. T.H. Chen, W.L. Lin, C.M. Liaw, Dynamic modelling and controller design of Flyback converter. IEEE Trans. Aerosp. Electron. Syst. **35**(4) (1999). https://doi.org/10.1109/7.805441
3. J.-W. Yang, H.-L. Do, Soft-switching dual-flyback dc–dc converter with improved efficiency and reduced output ripple current. IEEE Trans. Ind. Electron. **64**(5) 3587–3594 (2017). https://doi.org/10.1109/tie.2017.2652404
4. P.-L. Huang, D. Chen, C.-J. Chen, Y. Ming, An adaptive high-precision overpower protection scheme for primary-side controlled Flyback converters. IEEE Trans. Power Electron. **26**(10), 2817–2824 (2011). https://doi.org/10.1109/tpel.2011.2106223
5. Y. Liu, D. Zhang, Z. Li, Q. Huang, B. Li, M. Li, J. Liu, Calculation method of winding eddy-current losses for high-voltage direct current converter transformers. IET Electric Power Appl. **10**(6), 488–497 (2016). https://doi.org/10.1049/iet-epa.2015.0559
6. G. Chryssis, *High Frequency Switching Power Supplies: Theory and Design*, 2nd edn. (McGraw-Hill Publishing)
7. A.I. Pressman, *Switching Power Supply Design*, 2nd edn. (McGraw-Hill Publishing), pp. 209–222
8. K. Billing, T. Morey, *Switchmode Power Supply Handbook*, 3rd edn. (McGraw-Hill Publishing), pp. 3.139–3.350
9. R. Rashmi, A. Lembhe, P.A. Kharade, M.D. Uplane, Current controlled single-phase interleaved boost converter with power factor correction. Int. J. Adv. Res. Electr. Electron. Instrum. Eng. **5**(4) (2016). https://doi.org/10.15662/ijareeie.2016.0504055
10. L. Umanand, *Power Electronics Essentials and Application* (Wiley India Pvt. Ltd., 2013), pp. 549–560
11. R. Leeans, S.-H. Hsu, Design and implementation of self-oscillating flyback converter with efficiency enhancement mechanisms. IEEE Trans. Ind. Electron. **62**(11), 6956–6964 (2015). https://doi.org/10.1109/tie.2015.2436880

Performance Improvement of Light-Emitting Diodes with W-Shaped InGaN/GaN Multiple Quantum Wells

Himanshu Karan and Abhijit Biswas

Abstract Using APSYS simulation program, we investigate the optical performance of InGaN/GaN multiple quantum well (MQW) blue light-emitting diodes (LEDs) with a W-shaped well structure with respect to optical output power and internal quantum efficiency with variation in injection current. The concept of W-shaped quantum well is proposed to lower the polarization field and to obtain better overlapping between the peaks of electron and hole concentrations. Our proposed LED with W-shaped quantum wells exhibits 91% improvement in output power as compared to a rectangular quantum well LED. Furthermore, our proposed LED shows 20% efficiency drooping in contrast to 49% with a conventional LED at an input current = 120 mA. Moreover, we analyze our results with the help of band diagram, electron and hole concentrations, and also radiative recombination rate in the wells obtained from numerical simulation program.

Keywords APSYS simulation · Efficiency droop · LEDs · W-shaped InGaN MQWs

1 Introduction

Of late, blue light-emitting diodes (LEDs) using InGaN/GaN materials have gained significant importance for their tremendous commercial applications such as back-lighting unit for display, traffic signal, and also solid-state lighting [1–3]. But the internal quantum efficiency (IQE) of such blue LEDs is still low at large current densities, which is referred to as efficiency droop [4, 5]. Earlier investigations suggested some probable causes of efficiency droop like higher electron leakage [6], lower hole

H. Karan (✉) · A. Biswas
Institute of Radio Physics and Electronics, University of Calcutta,
92, Acharya Prafulla Chandra Road, Kolkata 700009, India
e-mail: himanshu.karan89@gmail.com

© Springer Nature Singapore Pte Ltd. 2017
J. Bhaumik et al. (eds.), *Communication, Devices, and Computing*, Lecture Notes in Electrical Engineering 470, https://doi.org/10.1007/978-981-10-8585-7_23

injection in active region [7], inhomogeneous electron and hole concentration profiles in multiple quantum wells (MQWs) [8], and Auger recombination [9]. Among them, higher electron leakage due to lower blocking capability and insufficient hole injection in the active region are marked to be main causes of efficiency droop. However, actual causes for efficiency drooping are still under debate. Some research groups are focused on to reduce the internal polarization field in the device such that the electron blocking will improve [10, 11]. To improve the performance of InGaN/GaN LEDs, many researchers employed different types of techniques like graded quantum barriers [12], AlGaN/GaN superlattice EBL [13], InGaN/GaN superlattice last quantum barrier [14], graded quantum wells [15], and double electron blocking layer [16]. Furthermore, the strong polarization field in the InGaN/GaN structure results in a larger separation between electron and hole concentration peaks which causes a reduction of the radiative recombination rate. Although extensive investigations have been carried out in order to increase output power and lower efficiency droop of blue light-emitting diodes particularly at high current densities, the desired target has not reached yet.

In order to alleviate the foregoing shortcomings, we propose an InGaN/GaN LED structure featuring W-shaped InGaN MQWs. The W-shaped QWs create reduced amount of internal polarization field, thereby diminishing the spatial separation between electron and hole peak concentrations, and improve radiative recombination rate. In this work, we present the optical performance of our proposed LED with reference to output power and efficiency droop and also compare them with the corresponding results pertaining to the conventional MQW LED.

2 Device Structure and Simulation Framework

InGaN/GaN multiple quantum well LED structures are fabricated on a c-plane sapphire substrate. A 50-nm-thick undoped GaN buffer layer is formed on the substrate. On top of the buffer layer a 4.5-μm-thick n-type GaN layer is grown epitaxially. The MQW active region is sandwiched between n-type GaN cathode layer with a doping concentration of 5×10^{18} cm^{-3} and a p-AlGaN electron blocking layer (EBL) as delineated in Fig. 1. The active region comprises five undoped InGaN quantum wells having thickness of 2 nm each and six undoped GaN barriers of 15 nm each. In the present work, we consider two MQW LED structures with different shapes of the quantum well, labeled as LED I and LED II. The LED I, a conventional structure reported by Tsai et al. [17], consists of rectangular quantum wells having thickness of 2 nm and In content of 0.21 (Fig. 2a). The LED II consists of quantum wells with W-shape of quantum wells in which In content varies from 21 to 16% over the

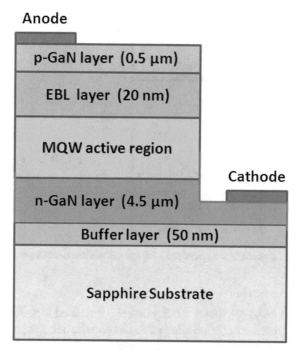

Fig. 1 Schematic structure of an InGaN/GaN multiple quantum well light-emitting diode

thickness of 1 nm, then 16 to 21% for the next 1 nm as illustrated in Fig. 2b. The EBL layer consists of 20-nm-thick p-type AlGaN having Al content of 0.15 and doping level of 1.2×10^{18} cm^{-3}. The whole device is capped by a 0.5-μm-thick p-type GaN layer having doping level of 1.2×10^{18} cm^{-3}. The ridge area of LED structure is 300×300 μm^2.

For numerical studies, we use APSYS simulation program [18] to obtain light–current (L–I) curve, IQE versus current, band diagram, carrier concentration profile, and radiative recombination rate for both conventional and our proposed LEDs. In the simulation for the quantum wells, we include Schrodinger and Poisson equations which are solved numerically to consider the quantum well deformation together with different bias conditions. We use the drift-diffusion carrier transport which includes continuity and current density equations of both electrons and holes. As the operation of LEDs relies on the spontaneous emission of photons due to electron–hole recombination in the quantum well, we use a free-carrier model taking into account the wurtzite valence band structure for the computation of optical power. Further details of the model can be found elsewhere [18]. The coefficients of Auger recombination, Shockley–Read–Hall (SRH) recombination, and spontaneous emission are used in our simulation to capture the emission and recombination events following Ref. [19]. The conduction-to-valence band offset ratio of the InGaN/GaN system is considered to be 70:30 [19]. We use energy gap for InN, GaN, and AlN in

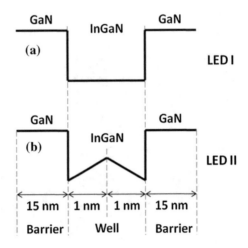

Fig. 2 **a** Rectangular quantum well profile of LED I and **b** W-shaped quantum well profile of LED II

accordance with Ref. [20]. The 6 × 6 k.p model is exploited for calculation of band structure in the presence of polarization field. The polarization charge density is computed using the approach followed by Fiorentini et al. [21]. It may be noted that we consider 0.4 fraction of the total polarization charge density in our simulation [19]. Our models used in the simulation program are calibrated by comparing simulated characteristics of optical output power and IQE against variation in input current with the corresponding experimental curve for LED I as reported in our earlier work [22].

3   Results and Discussion

Using APSYS software, we obtain optical output power and IQE with variation in input current for our proposed LED II. A comparison of output power as a function of injection current for both LED I and LED II is shown in Fig. 3. The figure shows a significant improvement (~91%) of output power in LED II compared to LED I. For instance, LED II yields an output power of 133.5 mW in contrast to 70 mW for LED I at injection current of 120 mA. Figure 4 shows dependence of the percentage of internal quantum efficiency with variation in input current. The values of efficiency droop, defined as the ratio of degraded efficiency to maximum efficiency at a higher current, are found to be 49% and 20% for LED I and LED II, respectively, at input current = 120 mA. In order to analyze the L–I and IQE versus current characteristics, we numerically obtain band diagram, carrier concentration profile, and radiative recombination rate for both LED I and LED II.

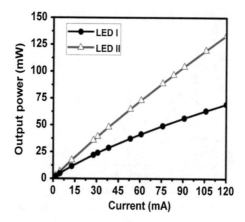

Fig. 3 Comparison of output power against input current for both LED I and LED II

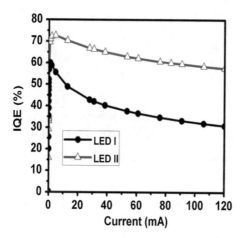

Fig. 4 Comparison of internal quantum efficiency against input current for both LED I and LED II

Figure 5 shows the band diagram of two LEDs along with quasi-Fermi levels for both electrons and holes at input current = 120 mA. In case of LED I, the InGaN quantum wells immediately follow GaN barrier producing larger lattice mismatch that develops higher strain in QWs. This higher strain creates the enhanced polarization field in the well, which in turn results in increased band bending in LED I shown in Fig. 5a. On the other hand, in a W-shaped QW the lattice mismatch is

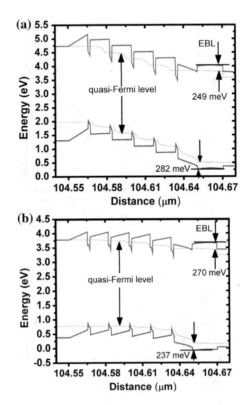

Fig. 5 Energy band diagram of **a** LED I and **b** LED II at injection current = 120 mA

reduced due to incorporation of In graded layers within the well and hence lesser band bending occurs in the QW of LED II compared to LED I as may be visualized in Fig. 5b. The effective barrier height values between the conduction band of EBL and the electron quasi-Fermi level are estimated to be 249 meV and 270 meV for LED I and LED II, respectively. Due to larger barrier height at EBL in LED II, the electron blocking capability in LED II is pretty higher than that of LED I. Figure 6a, b shows electron and hole concentrations in the quantum wells of LED I and LED II at injection current = 120 mA. The electron concentration increases in all QWs of LED II due to its stronger electron blocking capability than LED I as may be seen from Fig. 6a, b. Furthermore, the EBL in LED I provides an additional hole blocking potential which reduces injection of holes in the MQW region. On the contrary, the W-shape of quantum wells in LED II reduces the effective hole blocking potential at EBL compared to LED I. For example, the computed values of hole blocking potential between the valence band of EBL and hole quasi-Fermi level are 282 meV

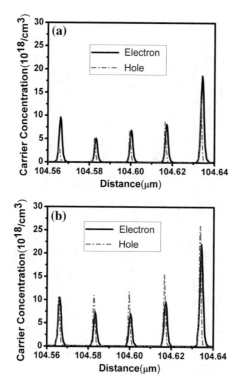

Fig. 6 Carrier distribution in the InGaN/GaN MQWs for **a** LED I and **b** LED II at injection current = 120 mA

and 237 meV for LED I and LED II, respectively, as indicated in Fig. 5a, b. It is evident from Fig. 6b that the concentration of holes is pretty higher together with more uniform distribution in all the wells of LED II relative to LED I. Figure 7a, b compares the concentration of electrons and holes with their peak separation in the fifth quantum well from cathode for LED I and LED II, respectively. The separation of 0.77 nm between the peak values of electron and hole concentrations in the well of LED I is higher than the corresponding separation of 0.52 nm in LED II. This feature is ascribed to the weak polarization field in the QW of LED II compared to LED I resulting in better overlapping of concentration peaks of electrons and holes in LED II.

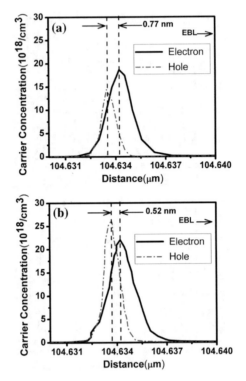

Fig. 7 Electron and hole concentrations in fifth quantum well from the cathode for **a** LED I and **b** LED II at input current = 120 mA

Figure 8a, b shows radiative recombination rate in all the QWs of LED I and LED II at input current = 120 mA. The radiative recombination rate is much higher in all QWs of LED II (Fig. 8b) than LED I (Fig. 8a) due to higher electron and hole concentrations and also their better overlapping. These features justify improved output power and reduced efficiency droop in LED II at higher values of injection current.

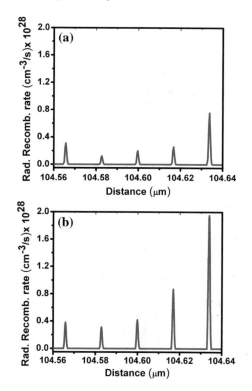

Fig. 8 Radiative recombination rate in **a** LED I and **b** LED II at input current = 120 mA

4 Conclusion

We have proposed and studied the optical performance of W-shaped InGaN/GaN multiple quantum well blue light-emitting diodes with respect to efficiency droop and optical output power. Our findings reveal that the proposed LED exhibits significantly higher optical output power and less amount of efficiency droop at high input current. We have also analyzed the operation of the proposed LED with the help of physical mechanisms. Hence, the proposed structure may be useful for constructing enhanced performance blue light-emitting diodes.

Acknowledgements The first author is thankful to UGC for supporting his fellowship vide no. F1-17.1/2013-14/RGNF-2013-14-SC-WES-52737/ (as may be seen in SA-III/Website).

References

1. C.-H. Cheng, A.-J. Tzou, J.-H. Chang, Y.-C. Chi, Y.-H. Lin, M.-H. Shih, C.-K. Lee, C.-I. Wu, H.-C. Kuo, C.-Y. Chang, G.-R. Lin, Growing GaN LEDs on amorphous SiC buffer with variable C/Si compositions. Sci. Rep. **6**, 19757–12 (2016)
2. X. Li, H. Sun, J. Cai, M. Yang, H. Zheng, H. Sun, Z. Guo, Study of blue InGaN/GaN light-emitting diodes with n-AlGaN layer as space layer and without the p-AlGaN electron blocking layer. J. Display. Technol. **11**(2), 170–174 (2015)
3. N. Tansu, H. Zhao, G. Liu, X.-H. Li, J. Zhang, H. Tong, Y.-K. Ee, III-nitride photonics. IEEE Photon. J. **2**(2), 241–248 (2010)
4. R.B. Chung, C. Han, C.C. Pan, N. Pfaff, J.S. Speck, S.P. DenBaars, S. Nakamura, The reduction of efficiency droop by $Al_{0.82}In_{0.18}N$/GaN superlattice electron blocking layer in (0001) oriented GaN-based light emitting diodes. Appl. Phys. Lett. **101**(13), 131113–3 (2012)
5. J. Piprek, Efficiency droop in nitride-based light-emitting diodes. Status Solidi A Phys. **207**(10), 2217–2225 (2010)
6. K.J. Vampola, M. Iza, S. Keller, S.P. DenBaars, S. Nakamura, Measurement of electron overflow in 450 nm InGaN light-emitting diode structures. Appl. Phys. Lett. **94**(6), 061116–3 (2009)
7. C.H. Wang, S.P. Chang, P.H. Ku, J.C. Li, Y.P. Lan, C.C. Lin, H.C. Yang, H.C. Kuo, T.C. Lu, S.C. Wang, C.Y. Chang, Hole transport improvement in InGaN/GaN light-emitting diodes by graded-composition multiple quantum barriers. Appl. Phys. Lett. **99**(17), 171106–3 (2011)
8. G.-B. Lin, D. Meyaard, J. Cho, E.F. Schubert, H. Shim, C. Sone, Analytic model for the efficiency droop in semiconductors with asymmetric carrier-transport properties based on drift-induced reduction of injection efficiency. Appl. Phys. Lett. **100**(16), 161106–4 (2012)
9. A. David, M.J. Grundmann, Droop in InGaN light-emitting diodes: A differential carrier life-time analysis. Appl. Phys. Lett. **96**(10), 103504–3 (2010)
10. C.H. Wang, C.C. Ke, C.Y. Lee, S.P. Chang, W.T. Chang, Hole injection and efficiency droop improvement in InGaN/GaN light-emitting diodes by band-engineered electron blocking layer. Appl. Phys. Lett. **97**(26), 261103–3 (2010)
11. S. Choi, M.-H. Ji, J. Kim, H.J. Kim, Md.M. Satter, P.D. Yoder, J.-H. Ryou, R.D. Dupuis, A.M. Fischer, F.A. Ponce, Efficiency droop due to electron spill-over and limited hole injection in III-nitride visible light-emitting diodes employing lattice-matched InAlN electron blocking layers. Appl. Phys. Lett. **101**(16), 161110–161114 (2012)
12. Z.G. Ju, W. Liu, Z.-H. Zhang, S.T. Tan, Y. Ji, Z.B. Kyaw, X.L. Zhang, S.P. Lu, Y.P. Zhang, B.B. Zhu, N. Hasanov, X.W. Sun, H.V. Demir, Improved hole distribution in InGaN/GaN light-emitting diodes with graded thickness quantum barriers. Appl. Phys. Lett. **102**(24), 243504–3 (2013)
13. J.H. Park, D.Y. Kim, S. Hwang, D. Meyaard, E.F. Schubert, Y.D. Han, J.W. Choi, J. Cho, J.K. Kim, Enhanced overall efficiency of GaInN-based light-emitting diodes with reduced efficiency droop by Al-composition-graded AlGaN/GaN superlattice electron blocking layer. Appl. Phys. Lett. **103**(6), 061104–4 (2013)
14. J. Chen, G.-H. Fan, G.-H. Pang, S.-W. Zheng, Y.-Y. Zhang, Improvement of efficiency droop in blue InGaN light-emitting diodes with p-InGaN/GaN superlattice last quantum barrier. IEEE Photon. Technol. Lett. **24**(24), 2218–2220 (2012)
15. A. Salhi, M. Alanzi, B. Alonazi, Effect of the quantum-well shape on the performance of InGaN-based light-emitting diodes emitting in the 400–500-nm range. J. Display Technol. **11**(3), 217–222 (2015)
16. Y. Guo, M. Liang, J. Fu, Z. Liu, X. Yi, J. Wang, G. Wang, J. Li, Enhancing the performance of blue GaN-based light emitting diodes with double electron blocking layers. AIP Adv. **5**, 037131–037136 (2015)
17. M.-C. Tsai, S.-H. Yen, Y.-C. Lu, Y.-K. Kuo, Numerical study of blue InGaN light-emitting diodes with varied barrier thicknesses. IEEE Photon. Technol. Lett. **23**(2), 76–78 (2011)
18. APSYS Software, (2015), www.crosslight.com

19. Y.-K. Kuo, M.-C. Tsai, S.-H. Yen, T.-C. Hsu, Y.-J. Shen, Effect of p-type last barrier on efficiency droop of blue InGaN light-emitting diodes. IEEE J. Quant. Electron. **46**(8), 1214–1220 (2010)
20. X. Yu, G. Fan, S. Zheng, B. Ding, T. Zhang, Performance of blue LEDs with n-AlGaN/n-GaN superlattice as electron-blocking layer. IEEE Photon. Technol. Lett. **26**(11), 1132–1135 (2014)
21. V. Fiorentini, F. Bernardini, O. Ambacher, Evidence for nonlinear macroscopic polarization in III–V nitride alloy heterostructures. Appl. Phys. Lett. **80**(7), 1204–1206 (2002)
22. H. Karan, A. Biswas, M. Saha, Improved performance of InGaN/GaN MQW LEDs with trapezoidal wells and gradually thinned barrier layers towards anode. Opt. Commun. **400**, 89–95 (2017)

Behavioral Modeling of Differential Inductive Seismic Sensor and Implementation of Its Readout Circuit

Abhishek Kumar Gond, Rajni Gupta, Samik Basu, Soumya Pandit and Soma Barman

Abstract Behavioral modeling of a new type of seismic sensor referred to as Differential Inductive Seismic (DIS) sensor for monitoring of seismic frequencies that are very close to the natural frequencies of buildings and other structures is presented in this paper. The model is validated by finite element method analysis using COMSOL Multiphysics 4.3a results show that the DIS sensor is highly selective to vibrations at resonance frequency and has good sensitivity and noise immunity. The sensor emulator and its readout circuit (ROC) are physically implemented on Texas Instruments (TI) ASLK PRO board. The ROC faithfully retrieves the single-tone seismic signal.

Keywords Seismic sensors · DIS sensor · Readout circuit · COMSOL

1 Introduction

Sensors used for measuring seismic vibrations are referred to as seismic sensors [1, 2]. At the present time, vibration sensors and accelerometers are widely used in earthquake detection and monitoring [3, 4]. The vibration sensor is categorized according to the sensing elements such as resistive, capacitive, inductive. Geophone is a classic example for this and is used as an earthquake detector for many years [5, 6]. But in recent years, it is being superseded by piezoelectric and variable capacitance transducers [7–9]. A notable feature in all these sensors is that they are inherently broadband sensors. This means that they will respond equally well to vibrations with a wide range of frequencies. However, in real life it is not necessary to evacuate a building or alarm its occupants as long as the frequencies of the vibrations are not sufficiently or nearly close to the natural frequency of the building, because the building is not in danger of collapsing. The current research works have mainly focused on the monitoring of seismic frequencies that are very close to or ideally

A. K. Gond · R. Gupta · S. Basu · S. Pandit · S. Barman (✉)
Institute of Radio Physics and Electronics, University of Calcutta, 92, A. P. C. Road, Kolkata 700009, India
e-mail: barmanmandal@gmail.com

© Springer Nature Singapore Pte Ltd. 2017
J. Bhaumik et al. (eds.), *Communication, Devices, and Computing*, Lecture Notes in Electrical Engineering 470, https://doi.org/10.1007/978-981-10-8585-7_24

equal to the natural frequencies of buildings and other structures, such as bridges [10–12]. Bakhoum et al. presented a frequency selective seismic (FSS) sensor based on the principle of electromagnetism, which is highly sensitive to vibrations at one specific frequency and inherently rejects vibrations at all other frequencies [3].

We propose here a differential inductive seismic (DIS) sensor which is improved alternative design of FSS sensor and has additional performance benefits. Due to differential inductive assembly, noise is low; SNR is very high and therefore has high selectivity. As this sensor possesses high sensitivity along with selectivity, so amplification is not required. The sensor can handle extremely low-frequency seismic signal (any specific frequency in the range of 0.2 to 4 Hz) which is excited by a high-frequency signal and also easy to process the signal. A comprehensive analytical behavioral model is presented here to design the sensor. The model is validated by finite element method analysis using COMSOL Multiphysics 4.3a. Various performance parameters, e.g., sensitivity, selectivity, SNR etc., have been studied for DIS sensor in COMSOL platform.

This paper is organized into five sections: Sect. 2 presents the working principle and mathematical formulation of the DIS sensor. Section 3 describes the implementation of readout circuit for the DIS sensor. In Sect. 4, we present results and discussions. Finally, we conclude the paper in Sect. 5.

2 Behavioral Modeling of Differential Inductive Seismic Sensor

2.1 Structure and Working Principle

A DIS sensor (Fig. 1) consists of two interdependent sections: first, a mechanical system comprising of a mass–spring–damper arrangement and second, an electrical system comprising of a differential coil transformer-type arrangement. The mass (m) is a ferrite core suspended by springs as shown in Fig. 1. It oscillates with respect to the coils, under the impact of seismic waves in forced harmonic oscillation of one degree of freedom (1 DOF). In absence of any seismic activity, the ferrite bead remains stationary at its mean position. So the two secondary coils are equally coupled with the primary. Equal voltages, but opposite in polarity, are generated in the two secondary coils. Hence, differential output from the secondary inductor is zero under balanced condition. The frequency of oscillation of the mass–spring system is tuned to the natural frequency of buildings, in the range of 0.2–4 Hz [10].

During earthquakes, the ground moves, together with the sensor assembly. However, the mass possessing higher inertia tends to remain stationary, but is forced to move under the action of the springs. One spring elongates, while the other compresses, exerting a balancing force to restore the core at null position. An excitation signal drives the primary coil located centrally and coaxially above the core.

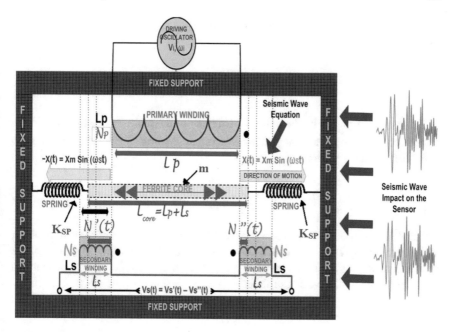

Fig. 1 Schematic diagram of a DIS sensor

The differentially connected secondary coils placed symmetrically on either side of the primary coil deliver the output.

As seismic vibrations impact the sensor, the mass/ferrite core exhibits motion. Thus, coupling between the secondary coils and the primary coil varies with time according to the instantaneous values of the seismic vibration. The output response of the DIS sensor resembles double-sideband suppressed-carrier modulated signal [13].

2.2 Mathematical Foundation

Mechanical system: To design the sensor, we require a seismic mass (m), a spring acting as a compliance (with spring constant K_{SP}), and a damper. Consider the seismic signal as an external driving force periodic in nature (function of time) acting on the mass. The force equation can then be written as

$$F = ma = m\frac{d^2x}{dt^2} = -bv - K_{SP}x + F_s \sin \omega_s t \tag{1}$$

where F_s is the seismic force, ω_s is the seismic frequency, m is the seismic mass, v is the seismic velocity, x is the seismic displacement, K_{SP} is the spring constant of the mass–spring system of the sensor, b = damping coefficient. Considering $b = 0$, we get

$$\frac{d^2x}{dt^2} + \frac{K_{SP}x}{m} = \frac{F_s}{m}\sin\omega_s t \tag{2}$$

Let $\omega_0 = \sqrt{K_{SP}/m}$ be the natural frequency of the mass–spring system, which is tuned to the natural frequency of the structure to be monitored. Then, the solution of the differential equation is

$$x(t) = \frac{(F_s/m)}{\omega_0^2 - \omega_s^2}\sin\omega_s t = X_s \sin\omega_s t \tag{3}$$

where X_s is the maximum displacement of the core at either side of the null position of the core during resonance, which occurs at $\omega_0 = \omega_s$. We denote (F_s/m) as a, where a is seismic acceleration. The value of a depends on the depth where the sensor has been placed below the earth's surface.

Electrical system: Let the input excitation to the primary coil be

$$V(t) = V_m \sin(\omega_i t) \tag{4}$$

where V_m is the amplitude of the input excitation signal and ω_i is the corresponding frequency. The current through the LR circuit formed between the inductance L_p of the primary coil and its internal resistance R_p is

$$i(t) = I_m \sin\omega_i t = \frac{V_m}{\sqrt{R_p^2 + (\omega_i L_p)^2}}\sin\omega_i t \tag{5}$$

A phase difference ϕ is generated between the voltage and current phasors in the LR circuit such that

$$V(t) = V_m \sin\left[\omega_i t + \tan^{-1}\left(\frac{\omega_i L_p}{R_p}\right)\right] \tag{6}$$

Let us suppose that ϕ_1 is the primary flux, ϕ_2 is the secondary flux, and $K' = \phi_{12}/\phi_1$ is the coefficient of magnetic coupling, where ϕ_{12} is the primary flux linked with the secondary. The maximum flux produced by the primary winding is

$$\phi_{PM} = \frac{L_p I_m}{N_p} = \frac{L_p V_m}{N_p\sqrt{R_p^2 + (\omega_i L_p)^2}} \tag{7}$$

Let the values of mutual inductance between the primary and the first secondary and between the primary and the second secondary be $M'(t)$ and $M''(t)$ respectively.

Likewise $N'_s(t)$ and $N''_s(t)$, be the effective number of turns getting coupled with the primary corresponding to $M'(t)$ and $M''(t)$ respectively.

$$N'_s(t) = n_s x(t) = (N_s/l_s) X_s \sin \omega_s t \tag{8a}$$

$$\text{and} \quad N''_s(t) = n_s [-x(t)] = - (N_s/l_s) X_s \sin \omega_s t \tag{8b}$$

where $n_s = N_s/l_s$ is turns per unit length of the secondary coil. Therefore, the mutual inductance becomes

$$M'(t) = \frac{N'_s(t)\phi_{12}}{i(t)} = \frac{N'_s(t)K'\phi_{PM}}{I_m} = \frac{N_s K' L_p}{l_s N_p} X_s \sin(\omega_s t) \tag{9a}$$

$$M''(t) = -\frac{N_s K' L_p}{l_s N_p} X_s \sin(\omega_s t) \tag{9b}$$

Now, the two secondary EMFs become

$$V'_s(t) = M'(t)\frac{di(t)}{dt} = \left[\frac{N_s K' L_p}{l_s N_p} X_s \sin(\omega_s t)\right] \frac{V_m \omega_i \cos(\omega_i t)}{\sqrt{R_p^2 + (\omega_i L_p)^2}} \tag{10a}$$

$$\text{Similarly,} \quad V''_s(t) = -\left[\frac{N_s K' L_p}{l_s N_p} X_s \sin(\omega_s t)\right] \frac{V_m \omega_i \cos(\omega_i t)}{\sqrt{R_p^2 + (\omega_i L_p)^2}} \tag{10b}$$

If the two secondary coils are connected in differential mode, then using (10a)–(10b), the resulting difference output voltage is given by

$$V_s(t) = \frac{K' K L_p V_m \omega_i X_s}{l_s \sqrt{R_p^2 + (\omega_i L_p)^2}} [\sin(\omega_i - \omega_s)t - \sin(\omega_i + \omega_s)t] \tag{11}$$

where $K = N_s/N_p$ is the turns ratio of the secondary to the primary winding.

It may be noted that (11) resembles a double-sideband suppressed-carrier (DSB-SC) amplitude-modulated signal (AM), where the carrier signal is the input excitation signal fed to the primary coil, and the modulating signal is the seismic mechanical vibration signal. The RMS value of the voltage generated at the output of the DIS sensor is given as

$$\left|V_{s(RMS)}\right| = \frac{K' K L_p V_m \omega_i}{l_s \sqrt{R_p^2 + (\omega_i L_p)^2}} \frac{a}{\left|\omega_0^2 - \omega_s^2\right|} \tag{12}$$

The signal-to-noise ratio in dB is given as

$$\left(\frac{S}{N}\right)_{dB} = 10\log_{10}\frac{\left|V_{s(RMS)}\right|^2}{\langle V^2 \rangle} = 10\log_{10}\frac{\left|V_{s(RMS)}\right|^2}{4kTBR} \tag{13}$$

where $\langle V^2 \rangle = 4\,kTBR$ is the thermal noise power in the coil, k is the Boltzmann's constant, T is the room temperature in kelvin, B is the bandwidth of the signal, and R is the total resistance of the windings.

3   Readout Circuit Design and Implementation

Due to complex nature of output waveform of the DIS sensor, we require a readout circuit (ROC) which gives a useful form of the output signal. The principle of operation of readout circuit for the DIS sensor is based on coherent demodulation technique [14]. The block diagram of the sensor emulator and ROC is shown in Fig. 2.

The complete circuit shown in Fig. 2 is implemented in hardware using Texas Instruments ASLK PRO board. To detect seismic signal, ROC requires the response of the DIS sensor. In our present work, we emulate the behavior of the DIS sensor using a circuit, shown in Fig. 3b referred to as sensor emulator. Here, signal 1 is the low-frequency seismic signal (10.5 Hz) and signal 2 refers to the input excitation signal (1 kHz), generated from sensor-driving circuit (Fig. 3a). The sensor emulator is implemented using IC MPY634, present on TI ASLK PRO board. In order to sustain the excitation signal of 1 kHz of driver circuit, two back-to-back diodes are used and gain needs to be greater than 3 [15]. This gain is achieved by the proper setting of 47 kΩ with 100 kΩ resistor. The output response of sensor emulator resembles like a double-sideband suppressed-carrier (DSB-SC) signal. The high-frequency rejection can be improved by trimming the offset voltage of the emulator circuit using 470 kΩ preset and 1 kΩ resistor. The DSB-SC signal, thus obtained, is fed to the detector circuit shown in Fig. 3c along with the same excitation signal. The output of the detector circuit is given to the Sallen–Key low-pass filter shown in Fig. 3d, which is designed for cutoff frequency (f_c) 10.5 Hz using R ≈ 16 kΩ and C = 1 μF. It removes the high-frequency components and also amplifies the signal at the output (V_{out}) of the filter by a gain of 11.

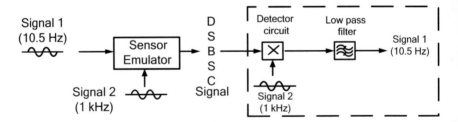

Fig. 2 Complete block diagram of sensor emulator and ROC

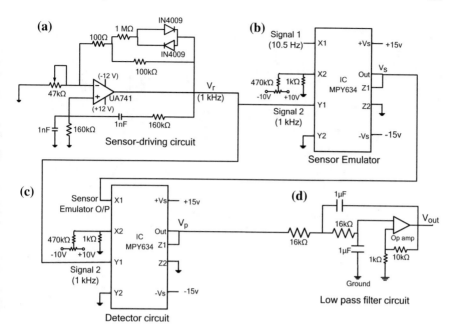

Fig. 3 Implementation of sensor emulator and ROC: **a** sensor-driving circuit, **b** sensor emulator circuit, **c** detector circuit, and **d** low-pass filter circuit

4 Results and Discussion

In order to validate the analytical model of the DIS sensor, we design the sensor structure (Fig. 4a) and simulate the response in COMSOL Multiphysics 4.3a environment using finite element method (FEM) analysis [16]. Several design parameters such as the magnitude of input voltage, coil excitation frequency, the number of turns in both coils, and the coil resistance are parameterized and can be changed as per requirements. The material of the core is chosen to be soft iron (without losses). The sensor structure is meshed using extremely fine tetrahedral elements to improve accuracy. Magnetic fields (mf) and electrical circuit (cir) physics are employed in solving the model. Gauge fixing is incorporated to numerically stabilize the model.

The variation of the output voltage generated by the DIS sensor with respect to time is shown in Fig. 4b. The simulation results show the small phase deviation between the analytical and FEM waveforms. This is due to some mismatch between the coefficient of coupling (K') used in analytical calculation and that used by the FEM simulation. Also, accuracy of the simulation result depends on the meshing strategy.

The variation of the RMS value of the EMF generated at the sensor output with respect to the displacement of the core is linear as shown in Fig. 5a. We observe that the value of the sensitivity of the DIS sensor is 39 mV/mm. Due to stray magnetic

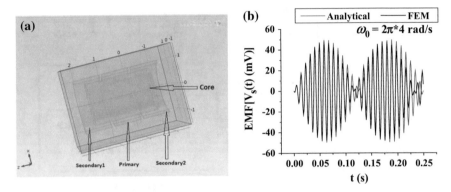

Fig. 4 a. Geometry design of the DIS sensor in COMSOL. **b** Output waveform of the sensor comparing results from theoretical data and data from FEM analysis

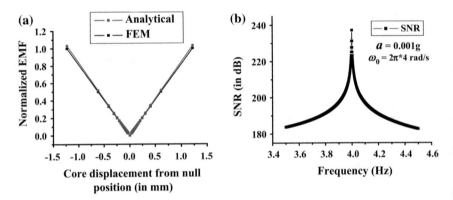

Fig. 5 a. Sensitivity plot of the DIS sensor for the case where the natural frequency of the oscillator is 4 Hz. **b** Plot of SNR versus frequency of the DIS sensor

field present in the system, a finite residual voltage, less than 1% (0.2 mV), is present at the null position of the core. Figure 5b shows the signal-to-noise ratio (SNR) vs frequency plot of the DIS sensor. This indicates that the DIS sensor has good noise immunity. Figure 5b also indicates the selective nature of the DIS sensor as the signal power is much strongly responding at the resonance frequency ($\omega_0 = \omega_s$) with minimum noise in it. In order to demonstrate the advantage of our DIS sensor, we compare the SNR response of our sensor with the FSS sensor, available in the literature [3] at a = 0.01 g. We find that the peak value of SNR at $\omega_s = 2\pi * 4$ rad/s is 260 dB, whereas that for the FSS sensor at a = 0.01 g and $\omega_s = 2\pi * 4$ rad/s is 90 dB.

In order to verify the ROC, we apply a single-tone seismic signal of frequency 10.5 Hz and excitation signal of frequency 1 kHz, as shown in Fig. 6a and 6b, respectively, to the sensor emulator circuit. Figure 6c shows the response of the

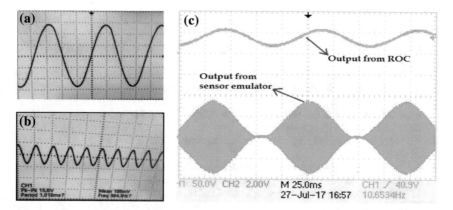

Fig. 6 **a** Low-frequency seismic signal (10.5 Hz) fed into the sensor emulator. **b** Sensor-driven signal (1 kHz) fed into the sensor emulator. **c** Output signal of the sensor emulator resembles a DSB-SC signal (below) and output signal of the ROC, retrieving the low-frequency signal of around 10.6 Hz (above)

emulator which is truly DSB-SC in nature. The ROC comprising of detector and filter faithfully reproduced the seismic signal frequency of 10.5 Hz shown in Fig. 6c.

5   Conclusion

This paper presents behavioral modeling of a DIS sensor, which is highly selective to vibrations at the resonance frequency which is tuned to the natural frequency of buildings and other structures to be monitored. The sensor structure is simulated in COMSOL Multiphysics 4.3a. The DIS sensor has higher value of peak SNR in comparison with the existing FSS sensor. The readout circuit along with a sensor emulator is implemented on ASLK PRO board, and the ROC faithfully reproduces the single-tone low-frequency signal.

Acknowledgements The authors thank SMDP-C2SD project, University of Calcutta, funded by DeitY, Govt. of India, for providing research facilities and fellowship, and also UPE II project, University of Calcutta, for partial financial assistance.

References

1. Institute of Mine Seismology. http://www.imseismology.org/seismic-sensors
2. R. PallÃis-Areny, J.G. Webster, Sensors and Signal Conditioning. 2nd edn., Wiley (2001)
3. E.G. Bakhoum, M.H.M. Cheng, Frequency selective seismic sensor. IEEE Trans. Instrum. Measure. **61**(3), 823–829 (2012)

4. L. Knopoff, Earthquake prediction: the scientific challenge. Proc. Nat. Acad. Sci. **93**(9), 3719–3720 (1996)
5. P. Gasparini, G. Manfredi, J. Zschau, *Earthquake Early Warning Systems* (Springer, Berlin, Germany, 2007)
6. C. Collette, P. Carmona-Fernandez, S. Janssens, K. Artoos, M. Guinchard, C. Hauviller, Review of sensors for low frequency seismic vibration measurement. CERN, ATS/Note/2011/001 (TECH)
7. F. Garcia, E.L. Hixson, C.I. Huerta, H. Orozco, Seismic accelerometer. in *Proceedings of IEEE IMTC*, pp. 1342–1347, May 1999
8. W. Boyes, *Instrumentation Reference Book* (Butterworh-Heinemann/Elsevier, Burlington, MA, 2010)
9. S.A. Dyer, *Survey of Instrumentation and Measurement* (Wiley, New York, 2001)
10. B.S. Smith, A. Coull, *Tall Building Structures* (Wiley, New York, 1991)
11. W.P. Jacobs, Building periods: moving forward and backward. Struct. Mag. **3**(6), 24–27 (2008)
12. K.M. Amanat, E. Hoque, A rationale for determining the natural period of RC building frames having infill. Eng. Struct. **28**(4), 495–502 (2006)
13. A.A. Spector, I.O. Martukhovich, Classification of Signals for Seismic Intrusion Alarm System
14. R. PallÃis-Areny, J. G. Webster, Analog Signal Processing. Wiley (1999)
15. D. Chattopadhyay, P.C. Rakshit, Electronics Fundamentals and Applications. NewAge International Publications (2008)
16. The COMSOL Multiphysics 4.3a User's Guide: 76-626, 940–1191

Application of PSO Variants for Optimal Design of Two-Stage CMOS Op-amp with Robust Bias Circuit

Bishnu Prasad De, K. B. Maji, Dibyendu Chowdhury, R. Kar, D. Mandal and S. P. Ghoshal

Abstract This paper investigates the relative optimizing proficiency between two PSO alternatives, particularly craziness-based PSO (CRPSO) and PSO with an aging leader and challengers (ALC-PSO) for the design of two-stage CMOS op-amp with robust bias circuit. PSO is a very simple optimization algorithm, and it copies the communal manner of bird flocking. The main disadvantages of PSO are premature convergence and stagnation problem. CRPSO and ALC-PSO techniques individually have eliminated the disadvantages of the PSO technique. In this paper, CRPSO and ALC-PSO are individually employed to optimize the sizes of the MOS transistors to reduce the overall area taken by the circuit while satisfying the design constraints. The results obtained individually from CRPSO and ALC-PSO techniques are validated in SPICE environment. SPICE-based simulation results justify that ALC-PSO is much better technique than CRPSO and other formerly reported method for the design of the aforementioned circuit in terms of the MOS area, gain, and power dissipation, etc.

Keywords Analog IC · CMOS two-stage op-amp · PSO · CRPSO · ALC-PSO
Robust bias

1 Introduction

Exact sizing of MOS transistors in VLSI circuit is a complex process. Evolutionary technique is a proficient option for the automation of sizing of MOS transistors in analog IC. Eberhart et al. [1] have developed the concept of PSO. Symmetric switching CMOS inverter using PSOCFIWA is reported in [2]. Area of CMOS oper-

B. P. De (✉) · D. Chowdhury
Department of ECE, HIT, Haldia, India
e-mail: bishnu.ece@gmail.com

K. B. Maji · R. Kar · D. Mandal
Department of ECE, NIT Durgapur, Durgapur, India

S. P. Ghoshal
Department of EE, NIT Durgapur, Durgapur, India

© Springer Nature Singapore Pte Ltd. 2017
J. Bhaumik et al. (eds.), *Communication, Devices, and Computing*, Lecture Notes in Electrical Engineering 470, https://doi.org/10.1007/978-981-10-8585-7_25

ational amplifier circuit [3, 4] is optimized by utilizing PSO technique. To obtain the maximum gain and UGB, folded cascode op-amp is designed by PSO method in [5].

Here, the area-optimized design of two-stage CMOS op-amp with robust bias circuit [6] is investigated by utilizing CRPSO [7] and ALC-PSO algorithm [8–10], individually.

The paper is written as follows: The steps for the design of the circuit are explained, and the cost function (CF) is described in Sect. 2. In Sect. 3, PSO, CRPSO, and ALC-PSO algorithms are described concisely. In Sect. 4, discussion of simulation results for the proposed techniques is given. Lastly, Sect. 5 concludes the paper.

2 Specifications for Design and Formulation of Cost Function

The circuit shown in Fig. 1 is optimally designed by using CRPSO and ALC-PSO, individually. The specifications for design are taken as follows: slew rate (SR), unity gain frequency (ω_u), minimum ICMR ($V_{IC\,(min)}$), maximum output voltage ($V_{out\,(max)}$), phase margin (ϕ_M).

The variables taken for the design are given as follows: width of channel (W) and length of channel (L) for the transistors present in the circuit, load capacitance (C_L). The steps for the design of the circuit [6] are given as follows:

Step 1: $C_C = \frac{16kT}{3\omega_u S_n(f)} \left[1 + \frac{SR}{\omega_u \left(V_{DD} - V_{IC}(\max) + V_{tn} \right)} \right]$

Step 2: $I_{D7} = SR\,(C_C + C_L)$

Step 3: $L_6 = \sqrt{\frac{3\mu_p \left(V_{DD} - V_{out}(\max) \right) C_C}{2\omega_u (C_C + C_L)\tan(\phi_M)}}$

Step 4: $W_6 = \frac{2SR(C_C + C_L)}{\mu_p C_{ox} \left(V_{DD} - V_{out}(\max) \right)^2} L_6$

Step 5: $I_{D5} = C_C(SR)$

Step 6: $\frac{W_1}{L_1} = \frac{W_2}{L_2} = \frac{\omega_u^2 C_C}{\mu_n C_{ox} SR}$

Fig. 1 Two-stage CMOS op-amp with robust bias circuit

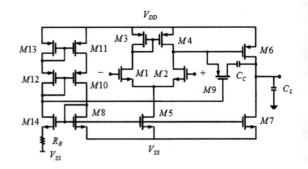

Step 7: $\dfrac{W_5}{L_5} = \dfrac{W_8}{L_8} = \dfrac{2(SR)C_C}{\mu_n C_{ox}\left(V_{IC(\min)} - V_{SS} - V_{tn} - \frac{SR}{\omega_u}\right)^2}$

Step 8: $\dfrac{W_7}{L_7} = \left(\dfrac{C_C + C_L}{C_C}\right)\left(\dfrac{W_5}{L_5}\right)$

Step 9: $\dfrac{W_3}{L_3} = \dfrac{W_4}{L_4} = \dfrac{\left(\frac{W_6}{L_6}\right)}{2\left(\frac{W_7}{L_7}\right)}\left(\dfrac{W_5}{L_5}\right)$

Step 10: $\dfrac{W_9}{L_9} = \left(\dfrac{C_C}{C_C + C_L}\right)\left(\dfrac{W_6}{L_6}\right)$

Step 11: $\dfrac{W_{10}}{L_{10}} = \dfrac{W_{11}}{L_{11}} = \dfrac{W_{12}}{L_{12}} = \dfrac{W_{13}}{L_{13}} = \dfrac{\left(\frac{W_6}{L_6}\right)}{\left(\frac{W_7}{L_7}\right)}\left(\dfrac{W_8}{L_8}\right)$

Step 12: $\dfrac{W_{14}}{L_{14}} = 4\left(\dfrac{W_8}{L_8}\right)$

Step 13: $R_B = \dfrac{1}{\sqrt{\left(2\mu_n C_{ox}\left(\frac{W_8}{L_8}\right)C_C(SR)\right)}}$

The size of initial population matrix is p × q, (p = 10 and q = 13). Rows (p) indicate the particle vectors present in the population. Columns (q) indicate the dimensions of particle vector and are denoted as

$$Y = [SR, \omega_u, V_{IC(\min)}, V_{out(\max)}, \phi_M, C_L, L_1, L_3, L_5,$$
$$L_7, L_9, L_{10}, L_{14}].$$

Therefore, the total number of variables to be optimized is q = 13. The area taken by all the transistors in the circuit is defined as cost function (CF) and is represented in (1).

$$CF = \sum_{i=1}^{14}(W_i \times L_i) \tag{1}$$

The number of transistors present in the circuit is 14. Here, CRPSO and ALC-PSO are individually applied for minimizing the CF.

3 Evolutionary Algorithms Used

In this section, PSO, CRPSO, and ALC-PSO are concisely presented. PSO is an optimization process dependent on population. PSO is a very simple method and is well explained in various literature [1, 2].

In birds' flocking, a bird habitually modifies directions unexpectedly. PSO is altered by presenting a different velocity equation correlated with various random numbers and the "craziness velocity." This altered PSO is called as CRPSO [7].

Table 1 Different parameters used for ALC-PSO algorithm

Parameters	CRPSO	ALC-PSO
Population size (p)	10	10
Dimension of the optimization problem (q)	13	13
Iteration cycle	100	100
C_1	2	2
C_2	2	2
P_{cr}	0.3	–
$v^{craziness}$	0.0001	–
Initial value of lifespan (Θ_0)	–	3

Idea of aging leader and challengers is introduced in the PSO to avoid premature convergence by producing challengers when current leader gets entrapped into local optima. ALC-PSO algorithm is well described in [8–10]. Parameters of CRPSO and ALC-PSO algorithm are provided in Table 1.

4   Discussions of Simulation Results

MATLAB is used to implement the CRPSO and ALC-PSO techniques to design the circuit (Fig. 1). The input variables are shown in Table 2. CRPSO and ALC-PSO are individually applied to attain the values of C_L, R_B, W_i, and L_i where i = 1, 2, ..., 14. For the authentication purpose, Cadence specter software (IC 5.1.41) is used to perform the transistor-level simulations of the circuit.

CRPSO and ALC-PSO are employed to design the circuit considering the constraints as

$$SR \geq 5\,V/\mu s, \ f_u \geq 5\,MHz, \ V_{IC(\min)} \geq -2.25, \ V_{out(\max)} \leq 2.2, \ \phi_M > 60^{\circ}$$

Table 2 Technology and inputs considered

Inputs, technology	Values taken
V_{DD} (V)	2.5
V_{SS} (V)	−2.5
V_{tp} (V)	−0.901
V_{tn} (V)	0.711
K_n' ($\mu A/V^2$)	182
K_p' ($\mu A/V^2$)	41.6
Technology	0.5 μm

Table 3 Constraints of the design parameters

Parameters	Ranges taken
SR (V/μs)	≥ 5
f_u (MHz)	≥ 5
$V_{IC(min)}$ (V)	≥ -2.25
$V_{out\,(max)}$ (V)	≤ 2.2
Phase margin (degree)	>60
C_L (pF)	≤ 10
C_C (pF)	0.5
L_i (μm)	$0.5\,\mu m \leq L_i \leq 5\,\mu m$
W_i (μm)	$0.75\,\mu m \leq W_i \leq 100\,\mu m$

Table 4 Design variables attained by different techniques

Design variables	[6]	CRPSO	ALC-PSO
W_1/L_1; W_2/L_2 (μm/μm)	1/1	1/1	1/1
W_3/L_3; W_4/L_4 (μm/μm)	1/1	1/4	1/3
W_5/L_5; W_8/L_8 (μm/μm)	1/3	1/2	1/2.2
W_6/L_6 (μm/μm)	31/2.5	9.99/2.5	8/2.25
W_7/L_7 (μm/μm)	3/1	5.7/1	4/1
W_9/L_9 (μm/μm)	1/1	1/2	1/2.2
W_{10}/L_{10}–W_{13}/L_{13} (μm/μm)	1/1	1/2	1/2.2
W_{14}/L_{14} (μm/μm)	2.5/2	1.23/1	1.2/1
C_c (pF)	0.5	0.5	0.5
C_L (pF)	5	4.5	4.25
R_B (kΩ)	Not reported	27.9	25

with the input values of $V_{DD} = 2.5\,V$, $V_{SS} = -2.5\,V$, $V_{tn} = 0.711\,V$, $V_{tp} = -0.901\,V$, $K'_n = 182\,\mu A/V^2$, $K'_p = 41.6\,\mu A/V^2$.

The design constraints are taken as $C_L \leq 10\,pF$, $C_C = 0.5\,pF$, $0.75\,\mu m \leq W_i \leq 100\,\mu m$, $0.5\,\mu m \leq L_i \leq 5\,\mu m$.

Process technology parameter used is 0.5 μm HP'S CMOS 14 TB [11]. The previously stated inputs and the constraints of the design parameters are given in Tables 2 and 3, respectively.

The optimal design variables attained by utilizing CRPSO and ALC-PSO, individually, are given in Table 4. Using the design variables, the circuit is re-designed in SPICE environment to obtain the specifications of the design.

Mahattanakul et al. [6] have designed the similar circuit studied in this paper. The simulation results reported in [6] have SR of 6.21 V/μs, f_u of 6.15 MHz, gain of 85 dB, and total transistor area of 100.5 μm^2.

Table 5 Comparison of performance specifications

Design criteria	Specifications	[6]	CRPSO	ALC-PSO
C_L (pF)	≤ 10	5	4.5	4.25
SR (V/µs)	≥ 5	6.21	5.2474	13.437
Power dissipation (µW)	<350	Not reported	347.5	342.5
Phase margin (°)	>60	65	68.56	66.56
Unity gain frequency (MHz)	≥ 5	6.15	11.2	12.2
Gain (dB)	>80	85	80.61	85.02
$V_{IC\ (min)}$ (V)	≥ -2.25	−2.2	−1.593	−1.615
$V_{IC\ (max)}$ (V)	≤ 2.25	2	1.474	1.732
$V_{out(min)}$ (V)	≥ -2.2	−2.15	−2	−2.1
$V_{out\ (max)}$ (V)	≤ 2.2	2.15	2	1.98
CMRR (dB)	>70	Not reported	90.36	96.36
$PSRR^+$ (dB)	>70	Not reported	110.1	84.58
$PSRR^-$ (dB)	>70	Not reported	83.86	106.9
Propagation delay (µs)	<1	Not reported	0.183	0.171
Input-referred noise @1 MHz ($nV/\sqrt{Hz}$)	< 45	44	39.85	38.85
Total MOS area (µm²)	<101	100.5	55.905	46.6

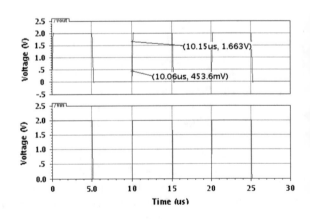

Fig. 2 Plot of SR

Fig. 3 Plot of power
dissipation

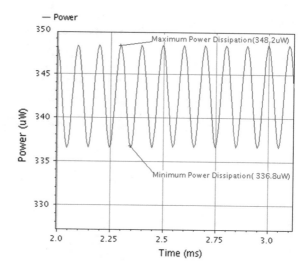

Fig. 4 Plot of UGB, gain,
and phase

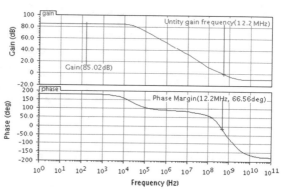

Fig. 5 Plot of ICMR

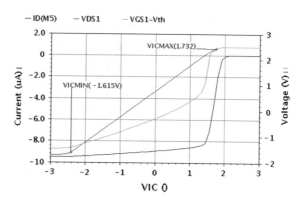

Fig. 6 Plot of output voltage swing

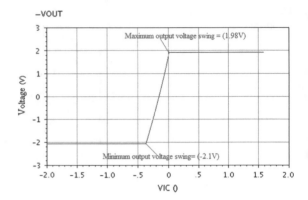

Fig. 7 Plot of CMRR

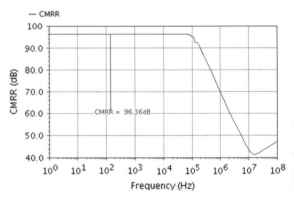

Fig. 8 Plot of positive PSRR

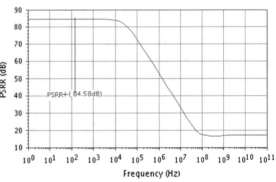

CRPSO method results in slew rate, unity gain frequency, gain, and total transistor area of 5.2474 V/μs, 11.2 MHz, 80.61 dB, and 55.905 μm^2, respectively, as given in Table 5.

ALC-PSO technique results in SR of 13.437 V/μs, f$_u$ of 12.2 MHz, gain of 85.02 dB, and total transistor area of 46.6 μm^2. So, ALC-PSO technique produces

Fig. 9 Plot of negative
PSRR

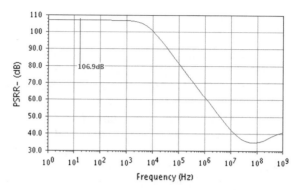

Fig. 10 Plot of delay

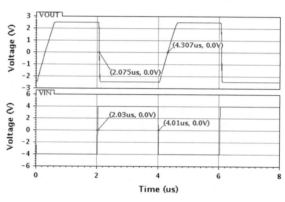

Fig. 11 Plot of noise

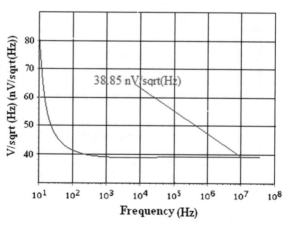

the most promising results for SR, UGB, gain, and the total transistor area. ALC-
PSO-based simulation results for the circuit obtained from SPICE are presented in
Figs. 2, 3, 4, 5, 6, 7, 8, 9, 10, and 11, respectively.

5 Conclusions

In this paper, two-stage CMOS op-amp with robust bias circuit is optimally designed by utilizing the CRPSO and ALC-PSO, individually. Both the CRPSO and ALC-PSO are proficient to generate the design optimal design variables. Simulations in SPICE environment have been executed by using the design variables obtained from CRPSO and ALC-PSO, individually. SPICE simulation results establish that evolutionary technique-based design meets all the specifications as well as reduce the total transistor area. ALC-PSO shows to be the best optimizer for the designed circuit under consideration. Future scope would be to explore the optimization efficiency of other metaheuristic methods for analog VLSI circuit sizing issues.

References

1. J. Kennedy, R. Eberhart, Particle swarm optimization, in *Proceedings of the IEEE International Conference on Neural Network*, vol. 4 (1995), pp. 1942–1948
2. B.P. De, R. Kar, D. Mandal, S.P. Ghoshal, Design of symmetric switching CMOS inverter using PSOCFIWA, in *Proceedings of the IEEE ICCSP' 14*, pp. 1818–1824, 3–5 April 2014, Melmaruvathur, Tamil Nadu, India
3. R.A. Vural, T. Yildirim, Analog circuit sizing via swarm intelligence. AEU Int. J. Electron. Commun. **66**(9), 732–740 (2012)
4. R.A. Vural, T. Yildirim, Swarm intelligence based sizing methodology for CMOS operational amplifier, in *Proceedings of the 12th IEEE Symposium on Computational Intelligence and Informatics* (2011), pp. 525–528
5. V. Ceperic, Z. Butkovic, A. Baric, Design and optimization of self-biased complementary folded cascode, in *Proceedings of the IEEE Mediterranean Electrotechnical Conference (MELECON)* (2006), pp. 145–148
6. J. Mahattanakul, J. Chutichatuporn, Design procedure for two-stage CMOS opamp with flexible noise power balancing scheme. IEEE Trans. Circuits Syst. **52**(8), 1508–1514 (2005)
7. B.P. De, R. Kar, D. Mandal, S.P. Ghoshal, Optimal analog active filter design using craziness based particle swarm optimization algorithm. Int. J. Numer. Model.: Electron. Netw. Devices Fields **28**(5), 593–609 (2015)
8. B.P. De, R. Kar, D. Mandal, S.P. Ghoshal, PSO with aging leader and challengers for optimal design of high speed symmetric switching CMOS inverter. Int. J. Mach. Learn. Cybern. **8**, 1403–1422 (2017)
9. B.P. De, R. Kar, D. Mandal, S.P. Ghoshal, An efficient design of CMOS comparator and folded cascode op-amp circuits using particle swarm optimization with an aging leader and challengers algorithm. Int. J. Mach. Learn. Cybern. **7**(2), 325–344 (2016)
10. B.P. De, R. Kar, D. Mandal, S.P. Ghoshal, Particle swarm optimization with aging leader and challengers for optimal design of analog active filters. Circuits Syst. Signal Process. **34**(3), 707–737 (2015)
11. R.J. Baker, H.W. Li, D.E. Boyce, *CMOS Circuit Design, Layout, and Simulation* (Wiley Interscience, New York, 1998)

Representation and Exploring the Semantic Organization of Bangla Word in the Mental Lexicon: Evidence from Cross-Modal Priming Experiments and Vector Space Model

Check for updates

Rakesh Dutta, Biswapati Jana and Mukta Majumder

Abstract In this paper, our primary intention is to determine the access mechanism and representation of semantically related Bangla word pairs in the mental lexicon. We conduct a visual priming experiment and user-annotated experiment over a number of native speakers in Bangla. After analyzing the response time and rate of errors, we observed that the priming is triggered to the semantically related word pairs in the mental lexicon. On the other hand, the response time data thus collected is used to evaluate on vector space model for finding the complete behavior of semantically related Bangla word pairs in the mental lexicon. The visual word recognition and interactive methods are to ensure with our result that the semantic priming may trigger or regulate the processing of word pairs at an early stage.

Keywords Semantic priming · Mental lexicon · Response time · Degree of priming · Vector space model

1 Introduction

Cognitive science is the study of human brain and its processes. The mental lexicon is the important part of cognitive science. It refers to the storage, organization, and connection of words in the human brain. These help to find the meaning and comprehension of words from the context. Bangla words are connected themselves in various ways (i.e., mainly in phonologically, morphologically, orthographically, and semantically) to represent the context. These connections are allowed to faster re-

R. Dutta (✉)
Department of Computer Science, Hijli College, Kharagpur 721306, West Bengal, India
e-mail: rakeshhijli@gmail.com

B. Jana
Department of Computer Science, Vidyasagar University, Midnapore 721102, West Bengal, India
e-mail: biswapatijana@gmail.com

M. Majumder
Department of Computer Science and application, University of North Bengal, Darjeeling 734014, West Bengal, India
e-mail: mukta_jgec_it_4@yahoo.co.in

© Springer Nature Singapore Pte Ltd. 2017
J. Bhaumik et al. (eds.), *Communication, Devices, and Computing*, Lecture Notes in Electrical Engineering 470, https://doi.org/10.1007/978-981-10-8585-7_26

trieval of words in the human brain, but these natures of words in the human brain are very much unknown to us. These are the areas of the current research. This research can help in several natural language processing applications like morphological segmentation, child education, machine translation, and information retrieval.

There is an abundant literature on the lexical access and organization of semantically related words. Many different operations have been processed in several languages such as English, Italian, Dutch, French, Hebrew, and few other languages [1–3]. However, no satisfactory investigations have been performed on Indian languages, where the accessing process of the semantically related word is completely different compared to its Indo-European cousins. The processing and mental representations of semantically related words are not language independent, which have been studied through several crosslinguistic experimentations [4]. Therefore, the experimental result in one language cannot be same to other languages. Hence, it has become necessary to perform same experiment in other languages.

In this paper, our goal is to find out whether the priming is active or not when the semantically related word appears. We have observed that a continuous effect occurs for the semantically related Bangla language. In short, highly similar words show a high degree of priming; commonly related words show intermediate priming, whereas less similar or dissimilar words show negative or less priming. Consequently, we have compared the priming observation with the computational model. Here, we have also found that "hyponym" relations show higher priming effects than "hypernym" relations.

The rest of the paper is organized as follows: Sect. 2 contains the related works. The cross-modal priming experiment 1 is given in Sects. 3 and 4 which contain user judgment priming experiment 2, and Sect. 5 contains the comparisons result with the result of a computational model (vector space model). The conclusion and discussion are presented in Sect. 6.

2 Related Work

One of the important studies in psycholinguistics shows the processing and representation of semantically related words in the human brain for different languages. It is called "semantic priming effect" [5, 6]. Early studies of semantic priming actually investigated that prime-target word pairs followed the norms of word association: A word such as "CAT" (the target) was identified significantly faster (in a lexical decision task) when it was appeared first by a related word (the prime, such as "DOG") than an unrelated word (such as "KEY") [7]. Undoubtedly, the effects of associative or semantic priming knowledge are necessary for understanding the access strategies and organization of the semantic or associative words in the mental lexicon.

Semantic relatedness refers to the similarity in meaning or the shared common features description of two words (e.g., "whale-dolphin"). Associative relatedness is a description of the probability that one word will help to fetch the second word in the human brain (e.g., "spider-web") [8]. The early studies on semantic or associative untangle relationship and the argument have been made in such a way that the associative relations do not arise from semantic similarity but rather from the co-occurrence of the word pairs in context [9, 10]. There are many examples of word pairs which are semantically related but not associatively related (e.g., truck-van). So, we have no proper definition of a "semantic relationship" [5, 11]. Clearly, an interesting issue is that the word pairs (i.e., prime and target) are semantically related but not associatively related will be the result of priming effect, again the word pairs (i.e., prime and target) are associatively related but not semantically related will also be the result of priming effect. So, the priming effect is the important issue to understand the lexical access and organization of semantically related word pairs in the human brain.

In this paper, we have focused on the study of semantic priming toward an Indian language (i.e., Bangla). We have performed two experiments and one comparison with the computational model to find out and prove the semantic similarities between word pairs in Bangla language. The first experiment has been made through cross-modal priming technique [DMDX]. The second experiment has been performed through the human annotation of hundred users. After that the results of two experiments were compared with the computational model [VSM], to find out the significant correlation of priming between two related Bangle words in the mental lexicon.

3 The Priming Experiment 1

In this experiment, we have chosen 300-word pairs for semantically measure and present the similar word pairs to 100 native Bangla Speakers. We have requested each user to provide a score for each word pairs in the range of 1 to 10 where ten implies highly semantically similar, and one means less semantically related. Based on the user-annotated score, we have been classified the word pairs and depicted in Table 1 below. We have conducted a visual priming experiment (i.e., cross-modal priming experiment) on the above user feedback dataset as presented in the next subsection. Here, a user visualizes a word on the display screen (called prime (P)) and immediately gets next word (called target (T)) from the word pairs. Based on the visual experiment, the users are asked to decide whether the target word is a valid word in a language (i.e., Bangla). The same experiment is also presented for same target word but a new word (called control (C)). The control words may not have any relationship with the targets. The time taken for lexical decision by the users after the

Table 1 User judgements of semantic similarity between numbers of words belonging to each class

Type of word	Percentage (%)
Highly similar	41
Average similar	52
Less similar	7

visual representation of the target is determined as the response time. The response time for both the experiments is analyzed and observed as the priming effect.

3.1 Materials

As discussed in the earlier section, the 300-word pairs are manually prepared from the Bangla corpus. The Bangla corpus of around 1 million words are already available.[1] From the 300-word pairs, we have classified the set of word pairs into two classes and have chosen 100-word pairs for each class. Class-I has represented the related word pairs, and Class-II has represented the unrelated word pairs. The example of the two classes is shown in the following Table 2. The motivation of choosing classes is to evaluate the priming effect. The Class-I consists related word pairs and to get the quick response for a priming effect. Class-II consists unrelated word pairs and to get the comparatively less response for a priming effect. The example of related word pairs is সূর্য (surya(sun)) and অস্ত (asta(sunset)) and unrelated word pairs is ছা-গল (chagal(goat)) and অস্ত (asta(sunset)).

In this experiment, the user performs the visual probe presentation. So, we put some fillers (or nonwords) to restrict the users to employ any strategically guesses during the experiments. Thus, in experiment 1, 100-word pairs for related class-I, 100-word pairs for unrelated class-II and 200-word pairs for nonwords (fillers); i.e., total 400-word pairs are sorted in a haphazard way to use as dataset. A set of 10 prime-target and control-target pairs were separately selected for an exercise session. However, the Response Time of these training pairs is excluded from any calculation or analysis.

3.2 Procedure

The experiment was conducted using graphical user interface software tool DMDX that displays the prime word (P), and then showed the target word (T) for 170 ms.[2] Corresponding to each visual probe, users had 2700 ms to perform the lexical decision

Table 2 Example of two classes for the experiment

Class	Explanation	Example
Class-I	Words are related	সূর্য (surya(sun))—অস্ত (asta(sunset))
Class-II	Words are not related	ছাগল (chagal(goat))—অস্ত (asta(sunset))

---

[1] http://www.isical.ac.in/~lru/downloadCorpus.html.
[2] http://linguistics.byu.edu/faculty/deddingt/604/DMDX.html.

before the system presents the next pairs of words. The user can decide the lexical task by stuffing either the "T" key for true word and "F" for false, respectively. The system keeps tracking the Response Time automatically.

Before starting the main experiment, we performed a trial test separately on ten-word pairs which were mentioned earlier. The experiment had two stages having same procedure, but word pairs were different. A break was given to the users about 2 hours to four days because each stage was sitting around 12 minutes having a lot of concentration. The break was categorized into three groups, i.e., 2 h, two days, and four days. From the results of experiments among three categories of rest, we observed that the better responses are coming from the users of four days rests than the 2 h rest. So, 2–4 days were given to the users for all the experiments. Again, each experimental stage was separated into four sub-settings of three minutes each and a break of two minutes was given between the settings.

3.3 Participants

The experiment was conducted on 100 native Bangla speakers among whom 50 users were appearing for the graduate degree, and 50 of them were from the twelve (X+II school) standards. The age of the users varies between 18 and 22 years.

3.4 Results and Discussion

Response Time less than the mean lexical latencies for correct response (i.e., 200 ms) or greater than the 2,000 ms and incorrect reaction (about 7.5%) was excluded from the latency analysis. The raw response time for all the correct responses was inversely transformed [12]. The mean response times for the prime-target and control-target sets are summarized in Table 3. From Table 3, we observe that the average response time for pairs of control-target was comparatively higher than that of the corresponding pairs of prime-target in Bangla.

Table 3 Average response time for the different classes

Class	Explanation	Average RT	Error (%)
Class-I	Words are related	676.8348808	3.2
Class-II	Words are not related	743.8018261	4.3

4 The Priming Experiment 2

4.1 Material and Procedure

In experiment 2, previous similar datasets are classified according to their semantic relationship. The datasets are grouped into three classes depending on the user ratings. The user's scores between (≥ 1 to ≤ 4) are in Class-I. The scores between (>4 to ≤ 7) are in Class-II, and the scores between (>7 to ≤ 10) are in third class. That is, the semantic scores are in a continuous value.

4.2 Participants

The experiment was conducted on 100 native Bangla speakers among whom 50 users were appearing for the graduate degree, and 50 of them were from the twelve (X+II school) standards. The age of the users varies between 18 to 22 years. None of them are allowed to take part in Experiment 1.

4.3 Results and Discussion

The judgment of this experiment is extremely user dependent where there is no control over users for executing such task. In order to obtain valid feedback from the users, we compute the Fleiss Kappa measure for finding the inter-annotator agreement [13]. The Kappa, k, can be defined as

$$k = \frac{\bar{P} - \bar{P}_e}{1 - \bar{P}_e} \tag{1}$$

The factor $\bar{P} - \bar{P}_e$ provides the degree of agreement that is achievable above chance, and, $1 - \bar{P}_e$ provides the degree of agreement truly attained above chance. We have obtained k = 0.58 which signifies that the user data are reliable. Now, we calculate the degree of priming for each class. The degree of priming means the average response of control-target word pairs minus the average response of prime-target words pairs. We have observed that degree of priming is depended on the semantic relationship. The degree of priming for each class is shown in Table 4. From Table 4, we observed that the highly semantically related words have a high degree of priming than the less semantically related word.

Table 4 The average response time depends on semantic priming score (given by user)

Group	Semantic distance	Degree of priming
1st group	≥ 1 to ≤ 4	37.44937
2nd group	> 4 to ≤ 7	38.62766
3rd group	> 7 to ≤ 10	50.98411

5 Analyzing the Effect of Semantic Priming Using Computational Model (VSM)

Based on the same dataset, we are calculating the semantic similarity on a vector space model to understand the lexical access and mental representation of word pairs in the human brain. We evaluate the outcomes of the model with the cross-modal priming experiments and the human judgment score. In the vector space model, the co-occurrence of words is observed from a large corpus and stored in a vector format. We calculate the co-occurrence of words using different approaches as suggested in the paper [14].

We have considered the co-occurrence as context words around the target word within a window of size 300 from the above-said Bangla corpus. The high-frequency content words are used to find out the co-occurrences from the corpus. Cosine similarity (sim) is used to calculate the similarity between two vectors. We have used raw co-occurrence value to construct the vector.

The semantically related word pair considered as W1, W2 (prime-target) and a non-semantically related word pair considered as W3, W2 (control-target). Sim are cosine similarity method between two word vectors computed as:

$$S1 = sim(v1, v2) = \frac{v1.v2}{|v1|.|v2|} \tag{2}$$

$$S2 = sim(v2, v3) = \frac{v2.v3}{|v2|.|v3|} \tag{3}$$

The cosine similarity function for the related class (Class-I) is sim (v1, v2) and nonrelated class (Class-II) is sim (v2, v3) where v1, v2, and v3 are the co-occurrence vectors of W1, W2, and W3 respectively.

In the final phase of our work, we observed that the semantic similarity for S1 is higher than the S2, shown in Table 5. The primary idea behind it is that the words are semantically similar if they are sharing number of common co-occurrences. Same types of observations are found for two classes. Overall, the results in Table 5 are agreed with that of Tables 3 and 4: The semantically related word pairs have degree of high priming than the non-semantic word pairs.

Table 5 Average cosine similarity value for different classes

Class	Explanation	Average of cosine similarity
Class-I	Words are related	$S1 = 0.945698607$
Class-II	Words are not related	$S2 = 0.544029$

6 Conclusions

In this paper, we have performed some basic psycholinguistic experiments to explore the representation and access mechanism of semantically related Bangla word pairs in the mental lexicon. For doing so, first we computed the visual priming experiment over the set of Bangla semantic word pairs. We observed that the highly semantically similar word pairs have significantly high priming; average semantically similar word pairs have intermediate priming and no or less priming for the low semantically related word pairs. On the other hand, we performed a user-annotated experiment to archive the priming effect for semantically related word pairs. Here we found the continuous effect of priming on semantically related word pairs.

In the next phase of our work, we examined the dataset collected from the user-annotated and visual priming experiments with the vector space model. However, it also concretes that the semantically related word pairs in Bangla have exhibited the priming behavior in the mental lexicon.

Despite the accuracy of the proposed model, it does not properly identify the synonym and polysemy words.

The semantic similarity-based approaches can be used for different purposes like categorizing the news or messages, document summarization, word sense disambiguation, and child education which we have considered as future scopes.

References

1. R. Frost, K.I. Forster, A. Deutsch, What can we learn from the morphology of Hebrew? A masked-priming investigation of morphological representation. J. Exp. Psychol. Learn. Mem. Cogn. **23**, 829–856 (1997)
2. J. Grainger, P. Cole, J. Segui, Masked morphological priming in visual word recognition. J. Mem. Lang. **30**, 370–384 (1991)
3. E. Drews, P. Zwitserlood, Morphological and orthographic similarity in visual word recognition. J. Exp. Psychol. Hum. Percept. Perform. **21**, 1098–1116 (1995)
4. M. Taft, Morphological decomposition and the reverse base frequency effect. Q. J. Exp. Psychol. **57A**, 745–765 (2004)
5. J.H. Neely, Semantic priming effects in visual word recognition: a selective review of current findings and theories, in *Basic Processes in Reading: Visual Word Recognition*, ed. by D. Besner, G.W. Humphreys. (Erlbaum, Hillsdale, NJ), pp. 264–336 (1991)
6. M. Lucas, Semantic priming without association: a meta-analytic review. Psychon. Bull. Rev. **7**, 618–630 (2000)

7. D.E. Meyer, R.W. Schvaneveldt, Facilitation in recognizing pairs of words: evidence of a dependence between retrieval operation. J. Exp. Psychol. **90**, 227–234 (1971)
8. L. Postman, G. Keppel, *Norms of Word Associations* (Academic Press, New York, 1970)
9. I. Fischler, Semantic facilitation without association in a lexical decision task. Mem. Cogn. **5**, 335–339 (1977)
10. J.A. Fodor, *Modularity of Mind* (MIT Press, Cambridge, MA, 1983)
11. K. McRae, V. de Sa, M.S. Seidenberg, On the nature and scope of featural representations of word meaning. J. Exp. Psychol. Gen. **126**, 99–130 (1997)
12. R. Ratcliff, Methods for dealing with reaction time outliers. Psychol. Bull. **114**(3), 510 (1993)
13. J.L. Fleiss, B. Levin, M.C. Paik, The measurement of interrater agreement. Stat. Methods Rates Prop. **2**, 212–236 (1981)
14. J. Mitchell, M. Lapata, Vector-based models of semantic composition, in *Proceedings of ACL-08: HLT*, pp. 236–244 (2008)

Solving a Solid Transportation Problems Through Fuzzy Ranking

Sharmistha Halder(Jana), Barun Das, Goutam Panigrahi and Manoranjan Maiti

Abstract A solid transportation problem (STP) with imprecise cost coefficients is modeled in this paper. The proposed fuzzy STP (FSTP) is optimized by the Vogel approximation method (VAM). For different types of imprecise operations and comparisons in VAM, fuzzy ranking method is used. The proposed FSTP is also converted to a crisp one using fuzzy expectation (mean value) and then solved using generalized reduced gradient (GRG) method. Finally, a numerical illustration is performed to support the methods, and optimum results by two methods are compared.

Keywords Solid transportation problem · Fuzzy rank · VAM · Fuzzy modi indices · Bounded technique

1 Introduction

Reality is less or more uncertain. Uncertainty can be represented and/or measured in different ways, like random, fuzzy, rough. In the last few decades, the fuzzy representation of uncertain parameters/variables has been done by most of the researchers (cf. Zimmermann [1], Liu [2], Lin and Liu [3], Maiti and Maiti [4]). Zadeh [5] first introduced the conception of fuzziness. There are various ways/processes of defuzzification like method of centroid, method of possibility. Ranking of a fuzzy number is an important feature to compare them. Ranking of fuzzy numbers is concerned with the

S. Halder(Jana) (✉)
Department of Mathematics, Midnapore College, Midnapore, India
e-mail: sharmistha792010@gmail.com

B. Das
Department of Mathematics, Sidho Kanho Birsha University, Purulia, India

G. Panigrahi
Department of Mathematics, NIT Durgapur, Durgapur, India

M. Maiti
Department of Mathematics, Vidyasagar University, Midnapore, India

© Springer Nature Singapore Pte Ltd. 2017
J. Bhaumik et al. (eds.), *Communication, Devices, and Computing*, Lecture Notes in Electrical Engineering 470, https://doi.org/10.1007/978-981-10-8585-7_27

defuzzification process also. Different ranking methods have been compared and reviewed by Bortolan and Degani [6] and more recently reviewed by Chen and Hwang [7], Choobineh and Li [8]. The alternative uses of fuzzy ranking numbers are studied by Dias [9], and automatic ranking of fuzzy numbers using artificial neural networks is proposed by Requena et al. [10]. Lee and Li [11], Raj and Kumar [12], Patra and Mondal [13] proposed a comparison of fuzzy numbers by considering the mean and dispersion (standard deviation) based on the uniform and proportional probability distributions.

A solid transportation problem (STP) is concerned with the selection of conveyance or path of a production–network–distribution system. It is a generalization of well-known classical transportation problem (TP). Hitchcock [14] first introduced the conception of solid transportation problem. Shell [20] stated the uses and abuses STP very carefully. Jiménez and Verdegay [18] investigated multi-objective STP in interval form and solved through genetic algorithm. Based on Chanas and Kuchta [15], Gen et al. [16], Ojha et al. [19] formulated an imprecise STP. Bit et al. [17] proposed an interactive fuzzy decision-making STP using linear and nonlinear membership function.

In this paper, a solid transportation problem is considered with imprecise costs. The proposed STP is solved using modified VAM. The obtained basic feasible solution is converted into an optimal one by MODI method(Method-1). The required operations and comparisons between the fuzzy numbers are executed by ranking method. The same FSTP is also solved by fuzzy expectation and generalized reduced gradient (GRG) methods using mean value and LINGO (12.0) respectively (Method-2). Finally, the proposed methods are illustrated with an example. The optimum results of two methods are compared, and it is seen that the minimum cost by VAM and MODI methods is less than the cost obtained by GRG method.

2 Mathematical Model of FSTP

In this section, a FSTP is mathematically modeled under the following assumption. Let us consider a STP with M origins $O_1, O_2, \ldots O_M$ of availability $a_1, a_2, \ldots a_M$ and N destinations $D_1, D_2, \ldots D_N$ of demand $b_1, b_2, \ldots b_N$. The items are transported through K conveyances with capacity $e_1, e_2, \ldots e_K$. If C_{ijk} be the unit transportation cost to transport the commodity from ith source to jth destination by kth conveyance, then the problem is to determine the quality x_{ijk} to be transported with respect to the total transportation cost subject to the availability constraints of the origins, demand constraints of the destinations, and capacity constraints of the conveyance. Hence, the mathematical model is given by:

$$\text{minimize} \quad \widetilde{Z} \;=\; \sum_{k=1}^{K}\sum_{i=1}^{M}\sum_{j=1}^{N} \widetilde{C}_{ijk}\, x_{ijk} \tag{1}$$

subject to the following constraints
Availability, demand, capacity, balance, and feasibility constraints are respectively.

$$\sum_{k=1}^{K}\sum_{j=1}^{N} x_{ijk} \leq a_i \quad \forall \ i = 1, 2, \dots M \tag{2}$$

$$\sum_{k=1}^{K}\sum_{i=1}^{M} x_{ijk} \geq b_j \quad \forall \ j = 1, 2, \dots N \tag{3}$$

$$\sum_{i=1}^{M}\sum_{j=1}^{N} x_{ijk} \leq e_k \quad \forall \ k = 1, 2, \dots, K \tag{4}$$

$$\sum_{i=1}^{M} a_i = \sum_{j=1}^{N} b_j = \sum_{k=1}^{K} e_j \tag{5}$$

$$x_{ijk} \geq 0 \quad \text{for all} \quad i, j, k. \tag{6}$$

3 Solution Method

3.1 Algorithm for Proposed Method-1

Step-1: First check the given STP is balanced or not. If not, change it into a balance one.
Step-2: Identify the cells having minimum and next to minimum transportation costs in each source and in each side of the table; write the difference (penalty) using the following definition
Definition: For a fuzzy number
$\tilde{A} = (a1, a2, a3, a4)$, $\overline{X}_{\tilde{A}} = \frac{1}{4}(a1 + a2 + a3 + a4)$, $S_{\tilde{A}} = (a4 - a1)$, $A_{\tilde{A}} = \frac{1}{2}(a4 + a3 - a1 - a2)$
The rank value of any fuzzy number (a1, a2, a3, a4) is $\leq a4$
proof: Consider $\tilde{A} = (a1, a2, a3, a4)$ as a trapezoidal fuzzy number where $a1 \leq a2 \leq a3 \leq a4$. Then, from the above definition:

The ranking value $R_{\tilde{A}} = \frac{1}{3}\{(2 \times \bar{X}_{\tilde{A}} + S_{\tilde{A}}) + A_{\tilde{A}}\}$
$= \frac{1}{3}\{2 \times \frac{1}{4}(a1 + a2 + a3 + a4) + (a4 - a1) + \frac{1}{2}(a4 + a3 - a1 - a2)\}$
$= \frac{1}{3}\{\frac{1}{2}(a1 + a2 + a3 + a4) + (a4 - a1) + \frac{1}{2}(a4 + a3 - a1 - a2)\}$
$= \frac{1}{3}\{2 * a4 + a3 - a1\}$
$\leq \frac{1}{3}(3 * a4)$ $[a3 - a1 \leq a4 - a1 \leq a4]$
$\leq a4 \, R_{\tilde{A}} = \frac{1}{3}\{(2 \times \bar{X}_{\tilde{A}} + S_{\tilde{A}}) + A_{\tilde{A}}\}$

Step-3: Apply the similar process step-2 for demands and conveyance (Identify the cells having minimum and next to minimum transportation costs in each demand and write the difference (penalty) using the definition of fuzzy ranking method, against the corresponding demand. Identify the cells having minimum and next to minimum transportation costs in each conveyance and write the difference (penalty) using following fuzzy rank method, against the corresponding conveyance).

Step-4: Identify the maximum penalty computed in step-3, make maximum allotment to the cell having minimum cost of transportation in that source or demand or conveyance. If in the table, two or more penalties are equal, there is a liberty to choose arbitrarily.

Step-5: Discard the fulfillment source or destination or conveyance corresponding to having maximum penalty and repeat the above steps (2–4) in the remaining problem until all restrictions are satisfied.

Step-6: The number of basic feasible solutions (obtained by followed steps 2–5) is (M+N+K−2). The problem is called degenerate if one of basic variable is zero, i.e., the number of nonzero basic variables is less than (M+N+K−2), and then, follow step-7; otherwise, it is called non-degenerate and goto step-8.

Step-7: For degenerate basic feasible solution, assign a small positive quantity $' \in'$ in a non allocated cell and goto step-8.

Step-8: Determine the set of (M+N+K) numbers MODI indices $\tilde{u}_i (i = 1, 2, \ldots, M)$, $\tilde{v}_j (j = 1, 2, \ldots, N)$ and $\tilde{w}_k (k = 1, 2, \ldots, K)$, in such a way that $\tilde{c}_{ijk} = \tilde{u}_i + \tilde{v}_j + \tilde{w}_k$, for all basic cells (i, j, k).

Step-9: For non-basic cell (i, j, k), we compute $\tilde{d}_{ijk}$ by using the formula $\tilde{d}_{ijk} = \tilde{c}_{ijk} - (\tilde{u}_i + \tilde{v}_j + \tilde{w}_k)$.

Step-10: Examine the matrix of cell evaluation $\tilde{d}_{ijk}$ for negative entries and conclude that

(i) If all $\tilde{d}_{ijk} > \tilde{0}$, then the solution is optimal and unique.
(ii) If all $\tilde{d}_{ijk} = \tilde{0}$, this imply the solution is optimal and alternate solution also exists.
(iii) If at least one $\tilde{d}_{ijk} < \tilde{0}$, then the solution is not optimal. Then, further improvement is necessary by following step-11 onwards.

Step-11:

(i) Find out the most negative cell in the matrix $[\tilde{d}_{ijk}]$.
(ii) Allocate θ to the empty cell in the final allocation table. Subtract and add the amount of this allocation to other sides of the loop in order to restore feasibility condition.

(iii) The value of θ is taken in such a way that no-one cell get negative allocation.

(iv) Put the value of θ and find a fresh allocation table.

Step-12: Again, apply the above test for optimality till you find all $\tilde{d}_{ijk} \geq \tilde{0}$

Optimality Condition Theorem

If $\{x_{ijk}^0, i = 1, 2, \ldots, M, \ j = 1, 2, \ldots, N, \ and \ k = 1, 2, \ldots, K\}$ is an optimal solution of the problem defined in Eqs. (1)–(6), then x_{ijk}^0 also gives the optimal solutions for the following problem.

$$\text{Minimize} \quad \tilde{Z} \ = \ \sum_{k=1}^{K} \sum_{i=1}^{M} \sum_{j=1}^{N} (\tilde{c}_{ijk} - \tilde{u}_i - \tilde{v}_j - \tilde{w}_k) \, x_{ijk} \tag{7}$$

subject to constraints

$$\sum_{k=1}^{K} \sum_{j=1}^{N} x_{ijk} \ = \ a_i \quad \forall \ i = 1, 2, \ldots M \tag{8}$$

$$\sum_{k=1}^{K} \sum_{i=1}^{M} x_{ijk} \ = \ b_j \ \forall \ j = 1, 2, \ldots N \tag{9}$$

$$\sum_{i=1}^{M} \sum_{j=1}^{N} x_{ijk} \ = \ e_k \quad \forall \ k = 1, 2, \ldots K \tag{10}$$

$$x_{ijk} \geq 0 \quad \text{for all} \ \ i, j, k.$$

$$(\tilde{c}_{ijk} - \tilde{u}_i - \tilde{v}_j - \tilde{w}_k) \geq 0 \quad \text{for all} \quad i, j, k.$$

where $\tilde{u}_i$, $\tilde{v}_j$, and $\tilde{w}_k$ are imprecise real values.

Proof.

As $\{x_{ijk}^0, i = 1, 2, \ldots, M, \ j = 1, 2, \ldots, N, \ and \ k = 1, 2, \ldots, K\}$ is an optimal solutions of the Eqs. (1)–(6), then clearly $\{x_{ijk}^0$ is a solutions of (7)–(10). Let if possible $\{x_{ijk}^0$ is not optimal. Then, there exists a feasible solution $\{x_{ijk}^0, i = 1, 2, \ldots, M, \ j = 1, 2, \ldots, N, \ and \ k = 1, 2, \ldots, K\}$ such that $\sum_{k=1}^{l} \sum_{i=1}^{m} \sum_{j=1}^{n}$ $c_{ijk} y_{ijk} < \sum_{k=1}^{l} \sum_{i=1}^{m} \sum_{j=1}^{n} c_{ijk} x_{ijk}^0$.

Clearly, $\{y_{ijk}, i = 1, 2, \ldots, m, \; j = 1, 2, \ldots, n, \; and \; k = 1, 2, \ldots, l\}$ is also a feasible solution of the problems (2)–(6).

Now

$$\widetilde{Z}_{min} = \sum_{k=1}^{K}\sum_{i=1}^{M}\sum_{j=1}^{N}(\widetilde{c}_{ijk} - \widetilde{u}_i - \widetilde{v}_j - \widetilde{w}_k) \; y_{ijk}$$

$$= \sum_{k=1}^{K}\sum_{i=1}^{M}\sum_{j=1}^{N}\widetilde{c}_{ijk}y_{ijk} - \sum_{k=1}^{K}\sum_{i=1}^{M}\sum_{j=1}^{N}\widetilde{u}_i\,y_{ijk} - \sum_{k=1}^{l}\sum_{i=1}^{M}\sum_{j=1}^{N}\widetilde{v}_j\,y_{ijk} - \sum_{k=1}^{K}\sum_{i=1}^{M}\sum_{j=1}^{N}\widetilde{w}_k\,y_{ijk}$$

$$< \sum_{k=1}^{K}\sum_{i=1}^{M}\sum_{j=1}^{N}\widetilde{c}_{ijk}y_{ijk} - \sum_{i=1}^{M}a_i\widetilde{u}_i - \sum_{j=1}^{N}b_j\widetilde{v}_j - \sum_{k=1}^{l}c_k\widetilde{w}_k$$

which contradicts $\{x_{ijk}^{0}, i = 1, 2, \ldots, M, \; j = 1, 2, \ldots, n, \; and \; k = 1, 2, \ldots, l\}$ is optimal solution of Eqs. (1)–(6). Therefore, $\{x_{ijk}^{0}, i = 1, 2, \ldots, M, j = 1, 2, \ldots, N \; and \; k = 1, 2, \ldots, K\}$ is an optimal solution of the problem (7)–(12).

3.2   Use of Fuzzy Expectation and GRG:

The fuzzy cost coefficients in (i) are transformed to crisp ones by fuzzy expectation, i.e., mean value for $\widetilde{A} = (a1, a2, a3)$. The reduced crisp objective function with crisp constraints is solved by generalized reduced gradient method using LINGO (12.0).

4   Numerical Experiments:

To illustrate the proposed model and verify the method, let us consider as numerical illustration of a STP with M = 3, N = 3, K = 3. A STP with three supplier three destinations and three conveyances in fuzzy environment is given in Table 1. The rank of each cost is calculated and given in Table 2. All allocations are calculated and given in Table 3. Calculated fuzzy numbers are given in Table 4.

4.1   Optimum Results by Method-1:

Consider a STP with three supplier three destinations and three conveyances in fuzzy environment. To compare the given fuzzy cost, rank of each cost are calculated.

Table 1 Suppliers with shipping cost

	D1			D2			D3			Availability
	E1	E2	E3	E1	E2	E3	E1	E2	E3	
O1	(2,4,4,8)	(5,7,11)	(2,5,12)	(1,2,7)	(6,8,14)	(2,7,7,9)	(1,5,5,9)	(2,6,11)	(1,2,3)	11
O2	(3,4,5)	(1,2,3)	(0,6,8)	(0,1,2)	(1,3,5)	(6,8,10)	(2,7,16)	(3,3,7)	(3,6,9)	13
O3	(2,5,5,12)	(0,1,2)	(2,3,4)	(2,4,6)	(5,7,9)	(3,3,3)	(4,5,7)	(5,6,8)	(3,4,5)	10
Capacity	11	14	9	11	14	9	11	14	9	
Demand	7			15			12			

Table 2 Rank is each cost

	D1			D2			D3			Availability
	E1	E2	E3	E1	E2	E3	E1	E2	E3	
O1	(2,4,4,8) R = 6.89	(5,7,11) R = 8.44	(2,5,12) R = 9.10	(1,2,7) R = 5.11	(6,8,14) R = 10.33	(2,7,7,9) R = 7.22	(1,5,5,9) R = 7.11	(2,6,11) R = 8.79	(1,2,3) R = 2.23	11
O2	(3,4,5) R = 3.89	(1,2,3) R = 2.44	(0,6,8) R = 7.10	(0,1,2) R = 1.11	(1,3,5) R = 4.33	(6,8,10) R = 7.11	(2,7,16) R = 12.51	(3,3,7) R = 4.79	(3,6,9) R = 7.23	13
O3	(2,5,5,12) R = 9.89	(0,1,2) R = 1.64	(2,3,4) R = 3.10	(2,4,6) R = 4.51	(5,7,9) R = 6.73	(3,3,3) R = 3.22	(4,5,7) R = 5.11	(5,6,8)	(3,4,5) R = 4.23	10
Capacity	11	14	9	11	14	9	11	14	9	
Demand	7			15			12			

Table 3 All calculated allocation

	D1			D2			D3			Availability
	E1	E2	E3	E1	E2	E3	E1	E2	E3	
O_1	0 (2,4,4,8)	0 (5,7,11)	0 (2,5,12)	0 (1,2,7)	0 (6,8,14)	0 (2,7,7,9)	0 (1,5,5,9)	2 (2,6,11)	9 (1,2,3)	11
O_2	0 (3,4,5)	0 (1,2,3)	0 (0,6,8)	11 (0,1,2)	2 (1,3,5)	0 (6,8,10)	0 (2,7,16)	0 (3,3,7)	0 (3,6,9)	13
O_3	0 (2,5,5,12)	0 (0,1,2)	7 (2,3,4)	0 (2,4,6)	0 (5,7,9)	0 (3,3,3)	0 (4,5,7)	1 (5,6,8)	0 (3,4,5)	10
Capacity	11	14	9	11	14	9	11	14	9	
Demand	7			15			12			

Following step-3, choose two minimum fuzzy cost for each origin, destination, and conveyance then calculate the corresponding penalties.

Along E_1, two minimum fuzzy costs are (0,1,2) and (3,4,5), and corresponding penalty is (1,3,5).

for O_1: (1,2,7)–(1,2,3) = (−2,0,6)
for O_2: (1,2,3)–(0,1,2) = (−1,1,3)
for O_3: (3,3,3)–(0,1,2) = (1,2,3)
for D_1: (1,2,3)–(0,1,2) = (−1,1,3)

Table 4 Calculated fuzzy numbers

	D_1			D_2			D_3			u_i
	E1	E2	E3	E1	E2	E3	E1	E2	E3	
O_1	[−16,4,28] (2,4,4,8)	[−14,7,24] (5,7,11)	[−21,12,42] (2,5,12)	[−19,−3,17] (1,2,7)	[−13,1,19] (6,8,14)	[−24,5,31] (2,7,9)	[−22,1,23] (1,5,5,9)	[−20,0,20] (2,6,11)	[−29,0,29] (1,2,3)	(−10,0,10)
O_2	[−4,9,22] (3,4,5)	[−5,5,15] (1,2,3)	[−14,13,37] (0,6,8)	[−10,0,10] (0,1,2)	[−8,0,8] (1,3,5)	[−11,9,30] (6,8,10)	[−11,7,28] (2,7,16)	[−9,1,14] (3,3,7)	[−17,8,33] (3,6,9)	(−8,−4,0)
O_3	[−4,10,20] (2,5,5,12)	[−6,0,6] (0,1,2)	[−2,6,25] (2,3,4)	[−8,−1,6] (2,4,6)	[−4,0,4] (5,7,9)	[−14,0,15] (3,3,3)	[−9,1,11] (4,5,7)	[−7,0,7] (5,6,8)	[−19,3,21] (3,4,5)	(0,0,0)
w_k	(0,5,10)	(5,7,9)	(−12,3,17)	(0,5,10)	(5,7,9)	(−12,3,17)	(0,5,10)	(5,7,9)	(−12,3,17)	
v_j	(−9,−6,−3)			(0,0,0)			(−4,−1,3)			

for D_2: $(3,3,3)-(0,1,2)=(1,2,3)$
for D_3: $(3,4,5)-(1,2,3)=(0,2,3)$
for E_1: $(0,1,2)-(3,4,5)=(1,3,5)$
for E_2: $(1,2,3)-(0,1,2)=(-1,1,3)$
for E_3: $(1,2,3)-(3,3,3)=(-2,-1,0)$

Select the maximum penalties by ranking method $(-2,0,6)$ corresponds to O_1 and followed step-5, allocate maximum possible amount to the cell $(1,3,3)$. And proceeding in this way, we get all the allocation as:

The number of B.F.S is $= (M + N + L - 2) = (3 + 3 + 3 - 2) = 7$, so the solution is non-degenerate. Then, we test the optimality by computing the MODI indices $\widetilde{u}_i, \widetilde{v}_j$ and $\widetilde{w}_k$ such that for all basic cells (i,j,k), $\widetilde{c}_{ijk} = (\widetilde{u}_i + \widetilde{v}_j + \widetilde{w}_k)$ and for non-basic cell (i,j,k) we compute $\widetilde{d}_{ijk}$; by using the formula $\widetilde{d}_{ijk} = \widetilde{c}_{ijk} - (\widetilde{u}_i + \widetilde{v}_j + \widetilde{w}_k)$, we get the following table.

4.2 Optimum Results by Method-2:

Finding the crisp mean values of fuzzy costs and then using the LINGO (12.0) software, the following optimum results are obtained.

Minimum cost $= 65$, optimum allocations are x221 $= 4$, x321 $= 7$, x112 $= 2$, x212 $= 5$, x222 $= 4$, x332 $= 3$, x133 $= 9$.

From this table, it is seen that the total transportation cost by Method-2 is higher than the cost obtained by Method-1.

5 Conclusion

This paper proposed a traditional solid transportation problem (STP) with imprecise cost parameters. The cost parameters are either triangular or trapezoidal fuzzy numbers. The fuzzy solid transportation problem (FSTP) is solved using Vogel's approximation method (VAM), where the different operations and comparisons are made using fuzzy ranking method. Moreover, the method and model are stabilized through a numerical example. The proposed fuzzy ranking method can be applied to other type of FSTP like FSTP with fixed charge, breakability, multi-item and also can be used to others decision-making problems like traveling salesman problem, inventory problem, and queening problem.

References

1. H.J. Zimmermann, Description and optimization of fuzzy system. Int. J. Gen. Syst. **2**, 209–215 (1976)

2. B. Liu, *Uncertainty Theory: An Introduction To Its Axiomatic Foundations* (Springer, Heidelberg, 2004)
3. L. Liu, L. Lin, Fuzzy fixed charge solid transportation problem and its algorithm. Fuzzy Syst. Knowl. Discov. **3**, 585–589 (2007)
4. A.K. Maiti, M. Maiti, Discounted Multi-item Inventory model via Genetic Algorithm with Roulette Wheel Section, arithmetic crossover and uniform mutation in constraints bounded domains. Int. J. Comput. Math. **85**(9), 1341–1353 (2008)
5. L.A. Zadeh, Fuzzy sets. Inf. Control **8**, 338–353 (1965)
6. G. Bortolan, R. Degani, A review of some methods for ranking fuzzy subsets. Fuzzy Sets Syst. **15**(1), 1–19 (1985)
7. S.J. Chen, C.L. Hwang, Fuzzy multiple attribute decision making methods, in *Fuzzy multiple attribute decision making* (Springer, Berlin, Heidelberg, 1992), pp. 289–486
8. F. Choobineh, H. Li, An index for ordering fuzzy numbers. Fuzzy Sets Syst. **54**(3), 287–294 (1993)
9. O.P. Dias, Ranking alternatives using fuzzy numbers: a computational approach. Fuzzy Sets Syst. **56**(2), 247–251 (1993)
10. I. Requena, M. Delgado, J.L. Verdegay, Automatic ranking of fuzzy numbers with the criterion of a decision-maker learnt by an artificial neural network. Fuzzy Sets Syst. **64**(1), 1–19 (1994)
11. E.S. Lee, R.J. Li, Comparison of fuzzy numbers based on the probability measure of fuzzy events. Comput. Math. Appl. **15**(10), 887–896 (1988)
12. P.A. Raj, D.N. Kumar, Ranking alternatives with fuzzy weights using maximizing set and minimizing set. Fuzzy Sets Syst. **105**(3), 365–375 (1999)
13. K. Patra, S.K. Mondal, Risk analysis in diabetes prediction based on a new approach of ranking of generalized trapezoidal fuzzy numbers. Cybern. Syst. **43**(8), 623–650 (2012)
14. F.L. Hitchcock, The distribution of a product from several sources to numerous localities. J. Math. Phys. **20**, 224–230 (1941)
15. S. Chanas, D. Kuchta, Multi-objective programming in optimization of interval objective functions—a generalized approach. Eur. J. Oper. Res. **94**, 594–598 (1996)
16. M. Gen, K. Ida, Y.Z. Li, E. Kubota, Solving bicriteria solid transportation problem with fuzzy numbers by a Genetic Algorithm. Comput. Ind. Eng. **29**, 537–541 (1995)
17. A.K. Bit, Fuzzy programming with hyperbolic membership functions for multi-objective capacitated solid transportation problem. J. Fuzzy Math. **13**(2), 373–385 (2005)
18. F. Jiménez, J.L. Verdegay, Interval multiobjective solid transportation problem via Genetic Algorithms. Manag. Uncertain. Knowl.-Based Syst. **II**, 787–792 (1996)
19. A. Ojha, B. Das, S. Mondal, M. Maiti, A solid transportation problem for an item with fixed charge, vechicle cost and price discounted varying charge using genetic algorithm. App. Soft Comput. **10**(1), 100–110 (2010)
20. E. Shell, Distribution of a product by several properties, Directorate of Management Analysis, in *Proceedings of the Second Symposium in Linear Programming*, vol. 2, (1995), pp. 615–642

Optimal Design of Low-Noise Three-Stage Op-amp Using PSO Algorithm

K. B. Maji, B. P. De, R. Kar, S. P. Ghoshal and D. Mandal

Abstract This paper adopts the bio-inspired Particle Swarm Optimization (PSO) algorithm for the design of a low-noise three-stage CMOS operational amplifier (TSCOA) circuit. The concept of PSO relies on the communal manner of bird flocking techniques. It is a very simple and easy to implement. The contribution of this work is to optimize the sizes of the individual MOS transistors by using PSO to reduce overall area of the circuit as well as the power dissipation. The optimized results are confirmed by Cadence simulator. The Cadence (IC 510) simulated results show that the design specifications are accurately met and the necessary functionalities are achieved. PSO-based design results in an improved functionality compared to those of the results reported in the recent literature.

Keywords TSCOA · Op-amp · CMOS · PSO · Optimization

1 Introduction

In VLSI domain, operational amplifier (op-amp) plays an important role. In most of the advance signal processing applications, op-amp is an important building block for amplification purposes. The design of an op-amp is tedious and time-consuming process. As the VLSI technology is scaled down, the CMOS circuit designer's face a lot of challenges to design an op-amp. Exact sizing of MOS transistors in VLSI circuit is a complex method. Evolutionary technique is a proficient option for automation in sizing of MOS transistors in analog IC. Kennedy and Eberhart [1] have developed the concept of PSO. An evolutionary approach-based design automation of low-power CMOS two-stage comparator and folded cascode by using

K. B. Maji (✉) · R. Kar · D. Mandal
Department of ECE, NIT Durgapur, Durgapur 713209, India
e-mail: kbmaji@gmail.com

S. P. Ghoshal
Department of EE, NIT Durgapur, Durgapur 713209, India

B. P. De
Department of ECE, HIT, Haldia, India

© Springer Nature Singapore Pte Ltd. 2017   293
J. Bhaumik et al. (eds.), *Communication, Devices, and Computing*, Lecture Notes
in Electrical Engineering 470, https://doi.org/10.1007/978-981-10-8585-7_28

Simplex-PSO was reported in [2]. Area of the CMOS operational amplifier circuit [3, 4] is optimized by utilizing PSO technique. To obtain maximum gain and UGB, folded cascode op-amp is designed by PSO method in [5].

In this paper, the area optimized design of CMOS three-stage op-amp circuit is investigated by utilizing the PSO algorithm.

The paper is structured as follows: The steps for the design of the circuit are explained, and the cost function (CF) is described in Sect. 2. In Sect. 3, PSO algorithm is described concisely. In Sect. 4, discussion of simulation results for the proposed technique is reported. Lastly, conclusion of the work is given in Sect. 5.

2 Optimization Algorithm Employed

Here, PSO algorithm is adopted for the optimal design of low-noise three-stage CMOS op-amp circuit.

Detailed illustrations of PSO can be found in [1]. In PSO, the particles' movements in the search space decide the particles' position and velocity. The particles' best position so far is denoted as P_{Best}. The best of P_{Best} values among all the particles is considered as G_{Best}. The velocity and position of each particle can be updated by (1) and (2), respectively.

$$V_{(t+1)} = W_t * S_t + C_1 * \text{rand}() * (P_{Best} - S_t) + C_2 * \text{rand}() * (G_{Best} - S_t) \tag{1}$$

$$S_{(t+1)} = S_t + V_{(t+1)} \tag{2}$$

where t is the iteration index; inertia weight is denoted by W; C_1 and C_2 are the acceleration constants; rand() is random function that lies in the range of [0, 1]. The new velocity of each particle is estimated by using (1). The particles' position is updated by using (2). Equations (3) and (4) are utilized for the calculation of C_1 and C_2, respectively.

$$C_1 = \text{C\_iter} * (C_{1e} - C_{1s})/\text{max\_iter} + C_{1s} \tag{3}$$

$$C_2 = \text{C\_iter} * (C_{2e} - C_{2s})/\text{max\_iter} + C_{2s} \tag{4}$$

where the initial values of C_1 and C_2 are denoted by C_{1s} and C_{2s}, respectively.

C_iter is the iteration number at the current generation, and max_iter is the maximum number of iteration used for the simulation. The final values of C_1 and C_2 are denoted by C_{1e} and C_{2e}, respectively. The pseudo-code for the implementation of PSO is given below.

Create and initialize the particles
Repeat until maximum iterations
    For each particle
        Estimate the fitness value
        If fitness value is better than P_{Best}
      Set fitness value as P_{Best}
    Select G_{Best} from P_{Best} of all particles
    Particle's velocity to be estimated by using (1)
    Particle's position is to updated by using (2)
End

3 Design Specifications for the CMOS TSCOA

The circuit considered in Fig. 1 is optimally designed by using PSO algorithm. Figure 2 represents a block diagram of the proposed circuit. The specifications for the design of the CMOS TSCOA are: power dissipation (P_{diss}), unity gain bandwidth (UGB), open-loop voltage gain (A_v), common-mode rejection ratio (CMRR), slew rate (SR). The variables adopted for the design are given as follows: length of the channel (L) and width of channel (W) for the transistors considered in the circuit and bias current (Ibias).

The open-loop gain or DC gain is calculated as $Gain(Av) = g1 * g2 * g3$, where, g1, g2, and g3 are the individual gain of the first, second, and third of the circuit, respectively.

$$|g1| = [g_{m2} * (r_{o2} \| r_{o4})] \tag{5}$$

$$|g2| = r_{o14} * \|[r_{o12} * (1 + g_{m12} * (r_{o10}(1 + g_{m10} \cdot r_{o6})))] \tag{6}$$

$$|g3| = [g_{m15} * (r_{o15} \| r_{o8})] \tag{7}$$

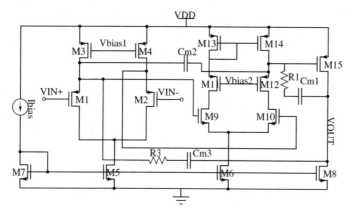

Fig. 1 Circuit diagram of the designed TSCOA

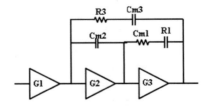

Fig. 2 Block diagram of the designed TSCOA

For PSO algorithm, the size of initial population matrix is m × n where (m = 10 and n = 13). Row (m) indicates the particle vectors present in the population. Column (n) indicates the dimensions of the particle vectors as given in (8).

$$X_{three-stage\_TSCOA} = [SR, A_v, UGB, CMRR, P_{diss}] \qquad (8)$$

The area occupied by all the transistors in the circuit is defined as the cost function which is presented in (9)

$$CF = \sum_{i=1}^{15} (W_i \times L_i) \qquad (9)$$

The number of transistors present in the circuit is 15, and the reported value of CF is given as 107.30 μm^2. Here PSO algorithm is applied to minimize the CF.

4 Discussions on Simulation Results

In order to validate the proposed TSCOA design, the transistors' dimensions of the op-amp are optimized by using PSO implemented in MATLAB R2016a version on CPU Intel core™ i5-vPro @3.00 GHz processor with 4 GB RAM and are verified by Cadence simulator (IC 5.1.41) in 0.18 μm technology. The supply voltage of the op-amp as 3.3 V is utilized here. The input variables are shown in Table 1.

Table 1 Inputs and technology considered for the design of TSCOA

Technology, inputs	Values taken
Technology	UMC 0.18 μm
V_{DD} (V)	1.8
K'_p ($\mu A/V^2$)	41.6
V_{tp} (V)	−0.901
V_{tn} (V)	0.711
K'_n ($\mu A/V^2$)	182

Table 2 Design constraints utilized

Parameters	Ranges taken
SR (V/μs)	≥5
Power (μw)	≥200
UGB (MHz)	≥5
Phase margin (degree)	>60
CMRR	≥90
L_i (μm)	$0.5\ \mu m \leq L_i \leq 5\ \mu m$
W_i (μm)	$0.75\ \mu m \leq W_i \leq 100\ \mu m$

Table 3 Control parameters used for PSO algorithm

Parameters	Value
Population size (p)	10
Dimension of the optimization problem (q)	13
Iteration cycle	200
C_1	2
C_2	2
Wt	0.95

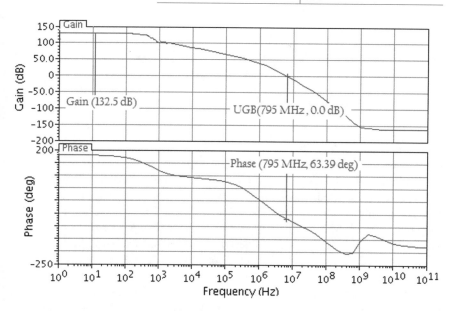

Fig. 3 Gain, phase, and UGB plot of TSCOA

The constraints of the design parameters are shown in Table 2. Different control parameters used for PSO algorithm are given in Table 3. The channel lengths for the transistors are taken as 2 μm [6]. The simulation results of gain, phase, and

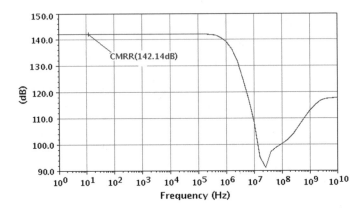

Fig. 4 Plot of CMRR of TSCOA

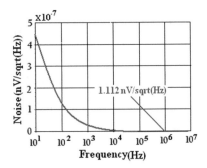

Fig. 5 Plot of input referred noise of TSCOA

unity gain bandwidth (UGB) are shown in Fig. 3. It is shown that open-loop gain or DC gain of the op-amp is 132.5 dB, unity gain bandwidth (UGB) is 795 MHz, and phase margin is 63.39°. The results presented in Fig. 3 confirm that the gain, phase, and UGB are individually much better than those of previous works [7–10]. The PM greater than 60° proves its stability. Figure 3 shows that the PM of the proposed design is 63.39° which is quite good for stability purpose. Thus, it proves that the proposed PSO-based circuit design is much more stable compared to those of the published works [7–10]. The plot of common-mode rejection ratio (CMRR) is shown in Fig. 4, and the value of CMRR has been achieved as 142.14 dB. The value of CMRR obtained is higher than those of the reported literature [7–10]. The average power dissipation is 160.995 μW, which is shown in Fig. 6. Plot of noise performance is presented in Fig. 5. It shows that the input referred noise is 1.112 nV/√Hz which is the lowest among those of the reported literature. Here, PSO technique results in SR of 13.437 V/μs as is presented in Fig. 7. PSO results in a total area of 107.08 μm² with the optimal design variables in 0.6085 s as shown in Fig. 8. The details of the simulated data for the TSCOA circuit are summarized in Table 4. So, from the

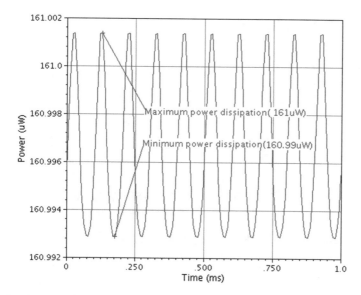

Fig. 6 Plot of average power dissipation of TSCOA

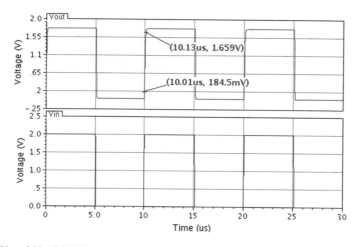

Fig. 7 Plot of SR of TSCOA

simulation results and Table 4, it is ensured that the optimal design of low-noise TSCOA is guaranteed.

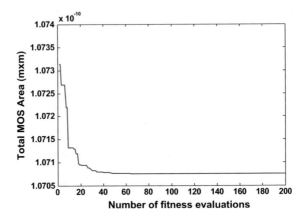

Fig. 8 Profile of convergence of PSO employed for TSCOA design

Table 4 Performance comparison

Performance	Present work	[7]	[8]	[9]	[10]
Technology (μm)	0.18	0.35	NR	0.35	0.18
Supply voltage V_{DD} (V)	3.3	3.3	± 2.5	3.3	3.3
Power (μW)	160.995	NR	NR	NR	NR
UGB (MHZ)	795	60	6.15	380	794
DC gain (dB)	132.5	110	85	110	128.5
Phase margin (deg)	63.39	60	65	10.6	59.5
CMRR (dB)	142.14	137.8	NR	NR	142
Output swing (V)	2.85	NR	NR	NR	2.77
Slew rate (SR) in V/μs	12.38	NR	NR	NR	NR
Area occupied by MOS transistor (μm^2)	107.30	NR	NR	NR	NR
Input referred noise $\left(nV/\sqrt{Hz}\right)$	1.112@ 1 MHz	24.92@ 1 MHz	44 @ 1 kHz	1.297@ 10 MHz	1.233@ 1 MHz

NR—Not referred in the literature

5 Conclusion

In this work, CMOS three-stage low-noise op-amp circuit is optimally designed by using PSO algorithm in UMC 0.18 μm process technology. The unity gain bandwidth is maximized to 795 MHz, and gain is 132.5 dB. The noise is minimized to 1.112 nV/$\sqrt{Hz}$. So, the proposed amplifier achieves the lowest noise performance and high gain simultaneously. SPICE-based simulation results established that PSO-based optimal circuit design achieves all the performance specifications compared to those of the other published literature. Simulation results also prove that PSO-based results provide better performance with respect to the previous works. Hence, the optimal design of three-stage low-noise op-amp circuit is guaranteed. As further work, these methods would be investigated in mixed-signal circuit optimization.

References

1. J. Kennedy, R. Eberhart, Particle swarm optimization, in *Proceedings of the IEEE International Conference on Neural Network*, vol. 4 (1995), pp. 1942–1948
2. K.B. Maji, R. Kar, D. Mandal, S.P. Ghoshal, An evolutionary approach based design automation of low power CMOS two-stage comparator and folded cascode OTA. AEU—Int. J. Electron. Commun. **70**(4), 398–408 (2016)
3. R.A. Vural, T. Yildirim, Analog circuit sizing via swarm intelligence. AEU Int. J. Electron. Commun. **66**(9), 732–740 (2012)
4. R.A. Vural, T. Yildirim, Swarm intelligence based sizing methodology for CMOS operational amplifier, in *Proceedings of the 12th IEEE Symposium on Computational Intelligence and Informatics*, Budapest, Hungary (2011), pp. 525–528
5. V. Ceperic, Z. Butkovic, A. Baric, Design and optimization of self-biased complementary folded cascode, in *Proceedings of the IEEE Mediterranean Electrotechnical Conference (MELECON)*(2006), pp. 145–148
6. P. Allen, D. Holberg, *CMOS Analog Circuit Design*, 2nd edn. (Oxford University Press, New York, 2002)
7. J.-L. Lai, T.-Y. Lin, C.-F., Yi-Te Lai and Rong-Jian, Design a Low-noise operational amplifier with constant-gm, in *SICE Annual Conference 2010*, pp. 322–326, Aug 2010, Taipei, Taiwan
8. J. Mahattanakul, J. Chutichatuporn, Design procedure for three-stage CMOS op-amp with flexible noise power balancing scheme. IEEE Trans. Circuits Syst. II: Express Briefs **52**(11), 766–770 (2005)
9. Z. Zhu, R. Tumati, S. Collins, R. Smith, D.E. Kotecki, A low-noise low-offset op-amp in 0.35 μm CMOS process, in *IEEE International Conference on Electronics, Circuits and system*(2006), pp. 625–627, Nice, France
10. A. Soltani, M. Yaghmaie, B. Razeghi, R. Pourandoost, S. Izadpanah Tous1, A. Golmakani, Three stage low noise operational amplifier design for a 0.18 μm CMOS process, in *International Conference on Electrical Engineering, Computing Science and Automatic Control (CCE)*, Mexico City, Mexico (2012), pp. 1–4

Optimal Design of Low-Voltage, Two-Stage CMOS Op-amp Using Evolutionary Techniques

Bishnu Prasad De, K. B. Maji, Banibrata Bag, Sayan Tripathi, R. Kar, D. Mandal and S. P. Ghoshal

Abstract This article explores the comparative optimizing efficiency of particle swarm optimization (PSO) and simplex-PSO method for the design of a low-voltage, two-stage CMOS op-amp. The concept of PSO is based on communal manner of bird flocking. The disadvantages of PSO are premature convergence and stagnation problem. Simplex-PSO is the combination of Nelder–Mead simplex method and PSO without considering the velocity term. The main idea is to optimize the size of the MOS transistors used for the op-amp circuit to reduce the overall area of the circuit. PSO- and simplex-PSO-based optimized results are confirmed by SPICE-based simulation. SPICE simulation results show that design specifications are approximately met and necessary functionalities are achieved. Simplex-PSO shows the better optimizing efficiency than PSO for the designed circuit.

Keywords Analog IC · Two-stage CMOS op-amp · Evolutionary optimization techniques · PSO · Simplex-PSO

1 Introduction

VLSI circuits in analog domain play a crucial character for the design of analog IC. Exact sizing of MOS transistors in VLSI circuit is a complex process. Evolutionary technique is a proficient option for the automation in sizing of MOS transistors

---

B. P. De (✉) · B. Bag · S. Tripathi
Department of ECE, HIT, Haldia, India
e-mail: bishnu.ece@gmail.com

K. B. Maji · R. Kar · D. Mandal
Department of ECE, NIT Durgapur, Durgapur, India

S. P. Ghoshal
Department of EE, NIT Durgapur, Durgapur, India

© Springer Nature Singapore Pte Ltd. 2017
J. Bhaumik et al. (eds.), *Communication, Devices, and Computing*, Lecture Notes in Electrical Engineering 470, https://doi.org/10.1007/978-981-10-8585-7_29

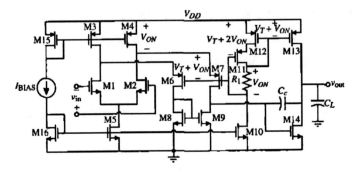

Fig. 1 A low-voltage, two-stage CMOS op-amp circuit

in analog IC. Eberhart et al. [1, 2] have developed the concept of PSO. Area of CMOS operational amplifier circuit [3, 4] is optimized by utilizing PSO technique. To obtain the maximum gain and UGB, folded cascode op-amp circuit is designed by PSO method in [5]. Simplex-PSO is utilized for the optimal selection of component values of analog active filers in [6]. Here, the area-optimized design of low-voltage, CMOS two-stage op-amp circuit is investigated by utilizing PSO and Simplex-PSO algorithms, individually.

The paper is written as follows: The design procedure of the circuit is explained, and the cost function (CF) is defined in Sect. 2. In Sect. 3, PSO and simplex-PSO algorithms are described concisely. In Sect. 4, discussion of simulation results is given. Lastly, conclusion of the work is given in Sect. 5.

2 Design Specifications and Objective Function Formulation

A low-voltage, two-stage CMOS op-amp circuit (Fig. 1) is optimally designed in this paper. Two-stage op-amp is a fundamental circuit used for the design of analog IC. Applications of this circuit are found in high gain amplification, filtering, analog-to-digital converters, and in many other systems.

The specifications for design of the circuit are taken as follows: slew rate (SR), output capacitance (C_L), unity gain bandwidth (UGB), maximum ICMR ($V_{IC\,(max)}$), and minimum ICMR ($V_{IC\,(min)}$). The variables taken for the design are given as follows: I_{BIAS}, width of channel (W) and length of channel (L) for the MOS transistors present in the circuit, C_L, compensation capacitance (C_C).

The steps for the design of the circuit [7] are given as follows:

- Supposing the condition to attain 60° phase margin and that the right half plane zero is as a minimum 10 GB gives

$$C_C \geq 0.2 C_L \tag{1}$$

- SR is linked to the I_{D5} as

$$I_{D5} = SR \cdot C_C \tag{2}$$

- Calculate the input transconductances of M1 and M2 from C_C and UGB.

$$g_{m1} = g_{m2} = GB \cdot C_C \tag{3}$$

The W/L ratios of M1 and M2 are expressed as

$$\frac{W_1}{L_1} = \frac{W_2}{L_2} = \frac{g_{m1}^2}{2K_n' \left(\frac{I_{D1}}{2}\right)} \tag{4}$$

- Calculate the W/L of M5 to validate

$$V_{IC(\min)} = V_{DS5}(sat) + V_{GS1} \tag{5}$$

$$V_{DS5}(sat) = V_{IC(\min)} - V_{GS1} = V_{IC(\min)} - \sqrt{\frac{2I_{D1}}{K_n'\left(\frac{W_1}{L_1}\right)}} - V_{tn} \tag{6}$$

$$\frac{W_5}{L_5} = \frac{2I_{D5}}{K_n' \left(V_{DS5}(sat)\right)^2} \tag{7}$$

- To design M3 and M4, the V_{IC} (max) is calculated as

$$V_{IC(\max)} = V_{DD} - V_{SD3}(sat) + V_{tn} \tag{8}$$

Knowing the current in M3, M4 gives

$$\frac{W_3}{L_3} = \frac{W_4}{L_4} = \frac{2I_{D3}}{K'_p \, (V_{SD3} \, (sat))^2} \tag{9}$$

- $V_{SD3} \, (sat) = V_{ON}$ of transistors M3 and M4, implement M10 through M12. Assume $I_{D10} = I_{D5}$, calculate the values of $\left(\frac{W_{10}}{L_{10}}\right)$ and R1. The W/L values of M11 and M12 are given as

$$\frac{W_{11}}{L_{11}} = \frac{W_{12}}{L_{12}} = \frac{2I_{D11}}{K'_p \, (V_{SD11} \, (sat))^2} \tag{10}$$

- W/L values of M6 and M7 are expressed as follows

$$\frac{W_6}{L_6} = \frac{W_7}{L_7} = \frac{W_{12}}{L_{12}} \tag{11}$$

- The gate–source voltage of M8 gives

$$V_{GS8} = V_{DD} - 2V_{ON} \tag{12}$$

Thus,

$$\frac{W_8}{L_8} = \frac{W_9}{L_9} = \frac{2I_{D8}}{K'_n \, (V_{DS8} \, (sat))^2} \tag{13}$$

- For 60° phase margin of $g_{m14} = 10g_{m1}$ and considering proper mirroring between M9 and M14, gives

$$\frac{W}{L} = \frac{g_m}{K'_n V_{DS} \, (sat)} \tag{14}$$

Calculate the value of $\left(\frac{W_{14}}{L_{14}}\right)$. I_{D14} is calculated by using this g_m and $\left(\frac{W_{14}}{L_{14}}\right)$. The $\left(\frac{W_{13}}{L_{13}}\right)$ is expressed as

$$\frac{W_{13}}{L_{13}} = \frac{I_{13}}{I_{12}} \frac{W_{12}}{L_{12}} \tag{15}$$

- To validate $V_{out \, (max)}$, the saturation voltage of M13 is

$$V_{SD13}(sat) = \sqrt{\frac{2I_{D13}}{K'_p \left(\frac{W_{13}}{L_{13}}\right)}} \tag{16}$$

- The W/L of M9 should be

$$\frac{W_9}{L_9} = \frac{I_{D9}}{I_{D14}} \frac{W_{14}}{L_{14}} \tag{17}$$

- For current mirror configuration of M3 and M15, gives

$$\frac{W_{16}}{L_{16}} = \frac{I_{BIAS}}{I_{D5}} \frac{W_5}{L_5} \tag{18}$$

- For current mirror configuration of M16 and M5, gives

$$\frac{W_{15}}{L_{15}} = \frac{I_{BIAS}}{I_{D3}} \frac{W_3}{L_3} \tag{19}$$

The cost function is given in (20).

$$CF_{op-amp} = \sum_{i=1}^{16} (W_i \times L_i) \tag{20}$$

The dimension of the optimization problem is 5 and defined as $X = [SR, C_L, UGB, V_{IC(min)}, V_{IC(max)}]$.

3 Evolutionary Algorithm Employed

PSO is a very simple method and is well explained in various literature [1, 2, 8]. The detailed description of simplex-PSO can be found [6]. Parameters of PSO and simplex-PSO algorithms are provided in Table 1.

Table 1 Different parameters for PSO and simplex-PSO algorithm

Parameters	PSO	Simplex-PSO
Population size (m)	10	10
Dimension of the optimization problem (n)	5	5
Iteration cycle	100	100
C_1	2	–
C_2	2	–
w	0.99	–
c_0	–	0.8
c_1	–	0.6
c_2	–	0.08

4 Simulation Results and Discussions

MATLAB is used to implement both the PSO and simplex-PSO techniques to design the circuit (Fig. 1). PSO and simplex-PSO are individually applied to attain the values of IBIAS, C_L, Cc, R1, and Wi (i = 1, 2, ..., 16). For the authentication purpose, Cadence (IC 5.1.41) is used to perform the transistor-level simulation of the circuit. Both the PSO- and simplex-PSO-based design results are described elaborately in this section. Both the PSO and simplex-PSO are used individually for the design op-amp circuit considering the design constraints of $SR \geq 10$ V/μs, $C_L \geq 10$ pF, $UGB \geq 10$ MHz, phase margin >60°, 0.5 V $\leq ICMR \leq 1.5$ V, $V_{out\,(max)} \geq 1.75$ V, $V_{out\,(min)} \geq 0.1$ V with inputs of $V_{DD} = 2$ V, $V_{SS} = -2$ V, $V_{tn} = 0.4761$ V, $V_{tp} = -0.6513$ V, $K'_n = 181.2$ μA/V^2, $K'_p = 65.8$ μA/V^2.

Process technology parameter used is 0.35 μm. The previously given constraints are presented in Tables 2 and 3. PSO results in a total transistor area of 173 μm^2 with optimal design parameters (I$_{BIAS}$, W$_i$, C$_L$, C$_c$, and R$_1$) in 3.746 s. Simplex-PSO results in a total transistor area of 108.5 μm^2 with particular values of optimal design parameters (I$_{BIAS}$, W$_i$, C$_L$, C$_c$, and R$_1$) in 3.148 s. The optimal design parameters attained from evolutionary techniques for the low-voltage, two-stage CMOS op-amp circuit are listed in Table 4. SPICE simulation results attained from PSO-based optimal design of the low-voltage, two-stage CMOS op-amp circuit are presented in Figs. 2, 3, 4, and 5, respectively. Figures 6, 7, 8, and 9 show the SPICE simulation

Table 2 Inputs and technology considered

Inputs, technology	Values considered
V_{DD} (V)	2
V_{SS} (V)	-2
V_{tp} (V)	-0.6513
V_{tn} (V)	0.4761
K'_n ($\mu A/V^2$)	181.2
K'_p ($\mu A/V^2$)	65.8
Technology	0.35 μm

Table 3 Constraints of the low-voltage, two-stage CMOS op-amp circuit

Parameters	Ranges considered
SR (V/μs)	≥ 10
C_L in (pF)	≥ 10
UGB in (MHz)	≥ 10
Phase margin (°)	>60
$V_{IC (min)}$ (V)	≥ 0.5
$V_{IC (max)}$ (V)	≥ 1.5
$V_{out (max)}$ (V)	≥ 1.75
$V_{out (min)}$ (V)	≥ 0.1
Length (L_i) in μm	1

results obtained from simplex-PSO-based optimal design of the circuit. The comparison of different performance parameters between PSO and simplex-PSO is shown in Table 5. Simplex-PSO-based optimal design technique for low-voltage, two-stage CMOS op-amp circuit offers much better results in terms of SR, gain, and total MOS area with respect to PSO-based design. So, the simplex-PSO demonstrates to be the best optimizer for this design. Layout diagram for simplex-PSO-based design of low-voltage, two-stage CMOS op-amp circuit is shown in Fig. 10.

Table 4 Design parameters obtained for low-voltage, two-stage CMOS op-amp circuit

Design variables	PSO	Simplex-PSO
I_{BIAS} (μA)	52	30
W_1/L_1; W_2/L_2 (μm/μm)	3.5/1	2/1
W_3/L_3; W_4/L_4 (μm/μm)	11.25/1	9.5/1
W_5/L_5; W_{10}/L_{10}; W_{16}/L_{16} (μm/μm)	8.5/1	9/1
W_6/L_6; W_7/L_7; W_{11}/L_{11}; W_{12}/L_{12}; W_{15}/L_{15} (μm/μm)	6.5/1	9/1
W_8/L_8; W_9/L_9 (μm/μm)	1/1	1/1
W_{13}/L_{13} (μm/μm)	62.5/1	3.5/1
W_{14}/L_{14} (μm/μm)	21/1	8/1
C_L (pF)	14.4	11.5
Cc (pF)	3.2	2.8
R_1 (k-ohm)	10.2	12.5

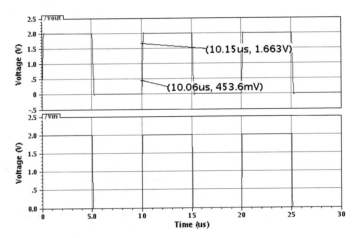

Fig. 2 Plot of slew rate of the PSO-based designed op-amp circuit

5 Conclusions

In this article, PSO and simplex-PSO methods are adopted for the optimal design of a low-voltage, two-stage CMOS op-amp circuit. Both the PSO and simplex-PSO are able to find out the optimal design parameters for the designed circuit. SPICE simulation has been executed by using the design parameters obtained from PSO and simplex-PSO individually. Simulation results establish that evolutionary technique-based design meets all the design specifications and also minimizes the total MOS transistor area. Simplex-PSO shows much improved results than PSO in terms of SR, gain, total MOS area, and execution time for the designed circuit.

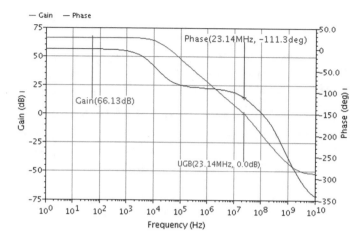

Fig. 3 Plots of UGB, gain, and phase of the PSO-based designed op-amp circuit

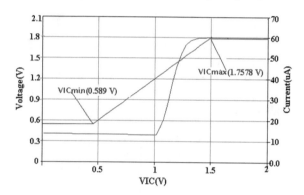

Fig. 4 Plot of ICMR of the PSO-based designed op-amp circuit

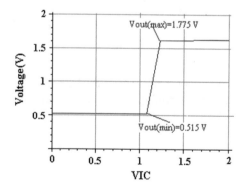

Fig. 5 Plot of output voltage swing of the PSO designed op-amp circuit

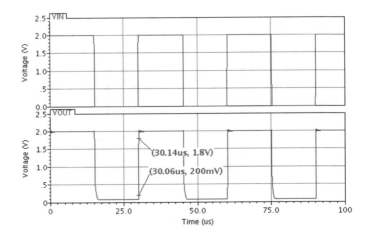

Fig. 6 Plot of slew rate of the simplex-PSO designed op-amp circuit

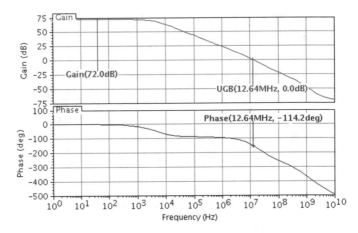

Fig. 7 Plots of unity gain bandwidth, gain, and phase of the simplex-PSO-based designed op-amp circuit

So, simplex-PSO is the better optimizer than PSO for the designed circuit. Future scope would be to explore the optimization ability of other metaheuristic methods for analog IC sizing issues.

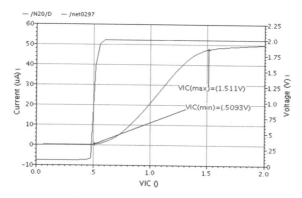

Fig. 8 Plots of ICMR of the simplex-PSO designed op-amp circuit

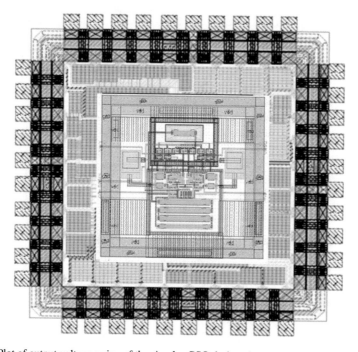

Fig. 9 Plot of output voltage swing of the simplex-PSO designed op-amp circuit

Table 5 Comparison of design specifications of low-voltage, two-stage CMOS op-amp circuit

Design Criteria	Specifications	PSO	Simplex-PSO
Slew rate (V/µs)	≥ 10	13.438	20
Phase margin (°)	>60	68.7	65.8
Unity gain bandwidth (MHz)	≥ 10	23.14	12.64
Gain (dB)	>60	66.13	72
$V_{IC\,(min)}$ (V)	≥ 0.5	0.589	0.5093
$V_{IC\,(max)}$ (V)	≥ 1.5	1.7578	1.511
$V_{out\,(min)}$ (V)	≥ 0.1	0.515	0.1106
$V_{out\,(max)}$ (V)	≥ 1.75	1.775	1.996
Total MOS area (μm^2)	<200	173	108.5

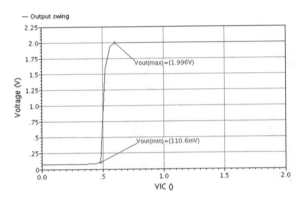

Fig. 10 Layout diagram of the simplex-PSO-based designed op-amp circuit

References

1. J. Kennedy, R. Eberhart, Particle swarm optimization, in *Proceedings of the IEEE International Conference on Neural Network*, vol. 4 (1995), pp. 1942–1948
2. R. Eberhart, Y. Shi, Comparison between genetic algorithm and particle swarm optimization, in *Evolutionary Programming-VII* (Springer, 1998), pp. 611–616
3. R.A. Vural, T. Yildirim, Analog circuit sizing via swarm intelligence. AEU Int. J. Electron. Commun. **66**(9), 732–740 (2012)
4. R.A. Vural, T. Yildirim, Swarm intelligence based sizing methodology for CMOS operational amplifier, in *Proceedings of the 12th IEEE Symposium on Computational Intelligence and Informatics* (2011), pp. 525–528
5. V. Ceperic, Z. Butkovic, A. Baric, Design and optimization of self-biased complementary folded cascode, in *Proceedings of the IEEE Mediterranean Electrotechnical Conference (MELECON)* (2006), pp. 145–148
6. B.P. De, R. Kar, D. Mandal, S.P. Ghoshal, Optimal selection of components value for analog active filter design using simplex particle swarm optimization. Int. J. Mach. Learn. Cybern. **6**(4), 621–636 (2015)
7. P. Allen, D. Holberg, *CMOS Analog Circuit Design*, 2nd edn. (Oxford University Press, New York, 2002)
8. B.P. De, R. Kar, D. Mandal, S.P. Ghoshal, Design of symmetric switching CMOS inverter using PSOCFIWA, in *Proceedings of the IEEE ICCSP' 14*, pp. 1818–1824, 3–5 Apr 2014, Melmaruvathur, Tamil Nadu, India

Printed in the United States
By Bookmasters